HEATH

ALGEBRA 2

AN INTEGRATED APPROACH

LARSON, KANOLD, STIFF

FORMAL ASSESSMENT

Norman B. Patterson

Evelyn A. Wedzikowski

International Standard Book Number: 0-395-87202-2

1 2 3 4 5 6 7 8 9 10 HWI 01 00 99 98 97

McDougal Littell

Evanston, Illinois • Boston • Dallas

TESTS

Mid-Chapter Tests

For each chapter in the Student Text, there are two Mid-Chapter Tests, Forms A and B, which are of average difficulty.

(In the Alternative Assessment booklet there are partner forms of the Mid-Chapter Tests which provide opportunities for cooperative learning.)

Chapter Tests

There are three Chapter Tests for each chapter in the Student Text. Forms A and B are of average difficulty and Form C is more challenging.

Cumulative Tests

Cumulative Tests are of average difficulty and should be used after every third chapter in the Student Text.

Chapters	Test	Use After	Form A	Form B	Form C
	Page		*Page*	*Page*	*Page*
1	Mid-Chapter Test	Lesson 1.4	1	2	---
	Chapter Test		3	6	9
2	Mid-Chapter Test	Lesson 2.4	12	13	---
	Chapter Test		14	17	20
3	Mid-Chapter Test	Lesson 3.3	23	24	---
	Chapter Test		25	28	31
1–3	Cumulative Test		34	---	---
4	Mid-Chapter Test	Lesson 4.4	42	43	---
	Chapter Test		44	48	52
5	Mid-Chapter Test	Lesson 5.3	56	57	---
	Chapter Test		58	62	66
6	Mid-Chapter Test	Lesson 6.3	70	71	---
	Chapter Test		72	76	80
1–6	Cumulative Test		84	---	---
7	Mid-Chapter Test	Lesson 7.3	92	93	---
	Chapter Test		94	97	100
8	Mid-Chapter Test	Lesson 8.3	103	104	---
	Chapter Test		105	108	111
9	Mid-Chapter Test	Lesson 9.4	114	115	---
	Chapter Test		116	119	122
7–9	Cumulative Test		125	---	---
10	Mid-Chapter Test	Lesson 10.3	133	134	---
	Chapter Test		135	139	143
11	Mid-Chapter Test	Lesson 11.3	147	148	---
	Chapter Test		149	152	155
12	Mid-Chapter Test	Lesson 12.3	158	159	---
	Chapter Test		160	163	166
7–12	Cumulative Test		169	---	---
13	Mid-Chapter Test	Lesson 13.3	177	178	---
	Chapter Test		179	182	185
14	Mid-Chapter Test	Lesson 14.3	188	189	---
	Chapter Test		190	193	196
15	Mid-Chapter Test	Lesson 15.3	199	200	---
	Chapter Test		201	204	207
13-15	Cumulative Test		210	---	---

Answers to Tests • Page 252

Preparation for College Entrance Tests Pages 218-251

To help students prepare for college entrance tests, we have provided strategies for taking the tests, instructions for using the new answer forms for student-generated responses, and practice questions.

Organization of Practice Questions

The practice questions are correlated to the chapters in the Student Text so that they require only information that has already been taught. The questions are also organized by type so that they provide organized practice in the three kinds of standardized test items: multiple choice, quantitative comparison, and student-generated response.

Form A
for each chapter:
multiple choice questions

Form B
for odd-numbered chapters:
quantitative comparisons

Form B
for even-numbered chapters:
student-generated responses

Practice for College Entrance Tests

ALTERNATIVE ASSESSMENT copymasters for *ALGEBRA 2* are also available from D.C. Heath. These include practical suggestions for alternative assessment activities, Chapter Projects, Math Logs, and Partner Quizzes.

Mid-Chapter 1 Test Form A

Name ______________________

Date ______________________

1. Plot the numbers on the real number line. Then decide which is greater.

$-\frac{1}{2}$ and $-\frac{5}{2}$

1. ______________ *Use graph at left.*

2. Write the numbers in *decreasing* order.

$7,\ -\sqrt{3},\ \frac{2}{3},\ 0,\ -\frac{7}{2},\ \frac{3}{5}$

2. ______________

3. What is the difference of 5 and -8?

3. ______________

4. Evaluate $(13 + 9x) \div 2y$ when $x = \frac{1}{3}$ and $y = -6$.

4. ______________

5. ***Geometry*** Find the area of the shaded region.

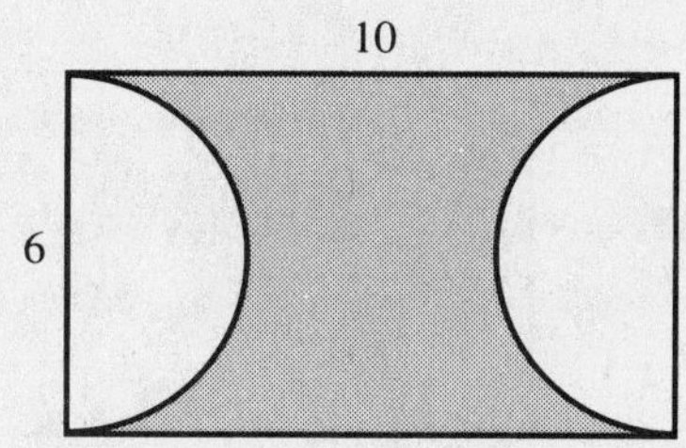

5. ______________

6. Evaluate the expression to two decimal places.

$15.72 - 3.87z$ when $z = 0.78$

6. ______________

7. Solve the equation. $-7 + 4x = 2$

7. ______________

8. Solve the equation. $-2(x - 3) + 4 = 4(2 - 3x)$

8. ______________

Mid-Chapter 1 Test Form B

Name ______________________

Date ______________________

1. Plot the numbers on the real number line. Then decide which is greater.

   -1 and $-\frac{7}{2}$

1. ______________ Use graph at left.

2. Write the numbers in *decreasing* order.

   $-5, \frac{2}{3}, -\frac{7}{4}, \frac{3}{2}, -\frac{5}{3}, \frac{5}{3}$

2. ______________

3. What is the difference of 6 and -4?

3. ______________

4. Evaluate $(13 + 9x) \div 2y$ when $x = -\frac{1}{3}$ and $y = 4$.

4. ______________

5. ***Geometry*** Find the area of the shaded region.

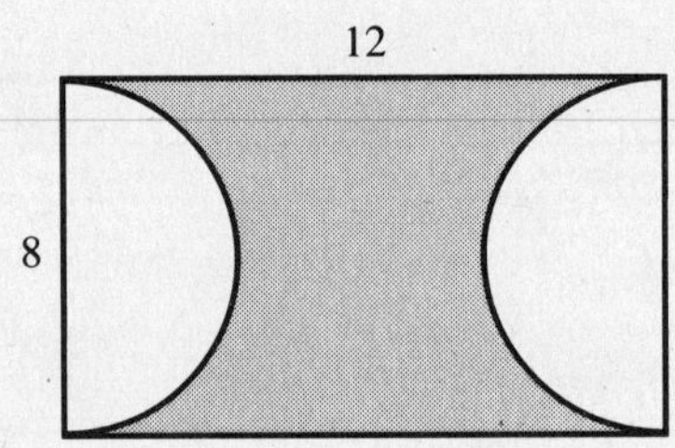

5. ______________

6. Evaluate the expression to two decimal places.

   $27.17 - 4.12z$ when $z = 2.84$

6. ______________

7. Solve the equation. $-6 + 3x = -20$

7. ______________

8. Solve the equation. $-3(x - 2) + 6 = 5(3 - 2x)$

8. ______________

Chapter 1 Test

Form A

(Page 1 of 3 pages)

Name ______________________

Date ______________________

1. Plot the numbers on the real number line. Then decide which is the greatest.

   $\frac{5}{2}, -\frac{3}{2}, 5$

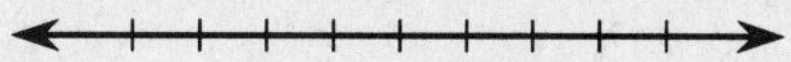

1. ______________ *Use graph at left.*

2. Write the numbers in *increasing* order.

   $\frac{5}{3}, -2, 0, -\frac{7}{2}, \frac{3}{5}, \frac{4}{3}, -1$

2. ______________

3. State the property that is illustrated.

   $3 \cdot (5 \cdot 7) = (3 \cdot 5) \cdot 7$

3. ______________

4. State the property that is illustrated.

   $5 + (-5) = 0$

4. ______________

5. What is the difference of 15 and -12?

5. ______________

6. Evaluate the expression. $80 - (20)(3) \div 5$

6. ______________

7. Evaluate $(7 + 5y) \div 3x$ when $x = \frac{1}{6}$ and $y = 3$.

7. ______________

8. Evaluate the expression to two decimal places.

   $37.15 - 4.55z$ when $z = 3.42$

8. ______________

Name ____________________

9. Solve the equation.

$-x + 3 = 7x + 8$

9. ____________________

10. Solve the equation.

$5(3 - 4x) = 7 - (4 - x)$

10. ____________________

11. ***Geometry*** Find the area of the shaded region.

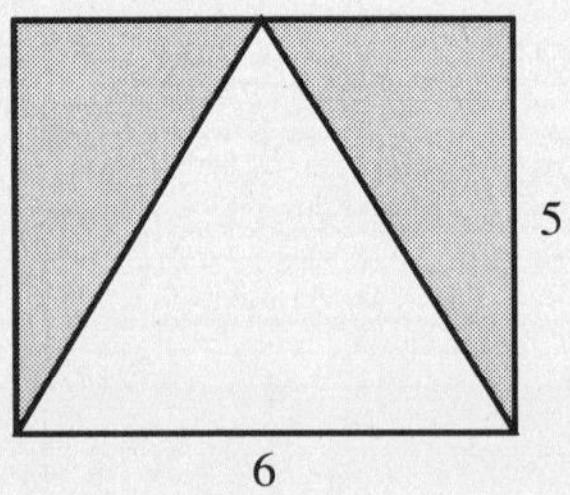

11. ____________________

12. Write the following expression using exponents.

"x cubed, times 3 to the nth power"

12. ____________________

13. Evaluate $2a^3 + (2a)^2$ when $a = -2$.

13. ____________________

14. ***Average Salary*** For 1980 through 1990, the average salary, A, (in 1000's of dollars), of assistant principals at public high schools can be modeled by $A = 2t + 25$ where $t = 0$ represents 1980. Approximate a high school assistant principal's salary in 1987.

14. ____________________

15. ***Temperature Conversion*** Solve for F.

$C = \frac{5}{9}(F - 32)$

15. ____________________

 Name ____________

16. Solve for q.

$p^2q - 3q = 14$

16. ____________

17. Solve the inequality and sketch its graph.

$x - 1 < -2(2 + x)$

17. ____________ *Use graph at left.*

18. Is $x = \frac{5}{2}$ a solution of the inequality

$5x - 4 \leq 3(x - 7)$?

18. ____________

19. Solve the inequality and sketch its graph.

$|3x - 2| \leq 5$

19. ____________ *Use graph at left.*

20. ***Stock Investment*** You have \$15,000 available to invest in two stocks, A and B. Write an inequality stating the restriction on A and B.

20. ____________

Chapter 1 Test Form B

(Page 1 of 3 pages)

Name ______________________

Date ______________________

1. Plot the numbers on the real number line. Then decide which is the greatest.

   $-\frac{7}{2}, \frac{1}{2}, 3$

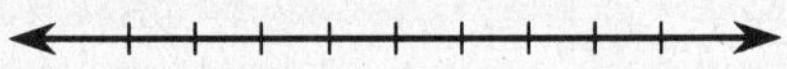

1. ______________ *Use graph at left.*

2. Write the numbers in *decreasing* order.

   $\frac{5}{3}, -2, 0, -\frac{7}{2}, \frac{3}{5}, \frac{4}{3}, -1$

2. ______________

3. State the property that is illustrated.

   $3 + (5 + 7) = (5 + 7) + 3$

3. ______________

4. State the property that is illustrated.

   $(a + 2b) + 3c = a + (2b + 3c)$

4. ______________

5. What is the difference of -7 and 15?

5. ______________

6. Evaluate the expression. $60 - (15)(2) \div 3$

6. ______________

7. Evaluate $(7 + 5y) \div 3x$ when $x = \frac{1}{9}$ and $y = 5$.

7. ______________

8. Evaluate the expression to two decimal places.

   $47.88 - 17.54z$ when $z = 1.44$

8. ______________

Name ______________________

9. Solve the equation.

$-3x + 5 = 7x + 8$

9. ______________

10. Solve the equation.

$4(2x - 3) = 6 - (3 - 2x)$

10. ______________

11. ***Geometry*** Find the area of the shaded region.

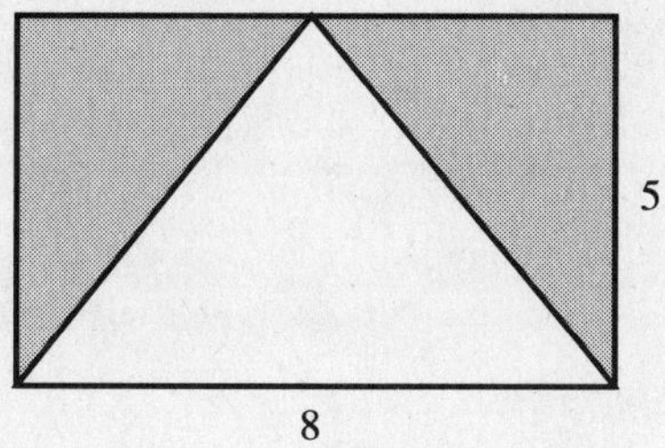

11. ______________

12. Write the following expression using exponents.

"3 cubed, times x to the nth power"

12. ______________

13. Evaluate $3a^3 + (3a)^2$ when $a = -3$.

13. ______________

14. ***Average Salary*** For 1980 through 1990, the average salary, A, (in 1000's of dollars), of assistant principals at public high schools can be modeled by $A = 2t + 25$ where $t = 0$ represents 1980. Approximate a high school assistant principal's salary in 1986.

14. ______________

15. ***Temperature Conversion*** Solve for C.

$F = \frac{9}{5}C + 32$

15. ______________

 Name ____________________

16. Solve for s.

$r^2s - 5s = 7$

16. ____________________

17. Solve the inequality and sketch its graph.

$2x + 5 \geq 2 - (x - 9)$

17. ____________________
Use graph at left.

18. Is $x = -\frac{5}{2}$ a solution of the inequality

$5x - 4 \leq 3(x - 7)$?

18. ____________________

19. Solve the inequality and sketch its graph.

$|2x - 3| \leq 5$

19. ____________________
Use graph at left.

20. ***Stock Investment*** You have \$25,000 available to invest in two stocks, A and B. Write an inequality stating the restriction on A and B.

20. ____________________

Form C

Name ______________________

Date ______________________

1. Plot the numbers on the real number line. Then decide which is the greatest.

$\frac{3}{5}, \frac{2}{3}, \frac{5}{3}$

1. ______________ Use graph at left.

2. Write the numbers in *increasing* order.

$\frac{3}{5}, 0, -\frac{1}{3}, \frac{2}{3}, -\frac{2}{3}, -1$

2. ______________

3. State the property that is illustrated.

$14(\frac{1}{14}) = 1$

3. ______________

4. State the property that is illustrated.

$2(x + y) + z = z + 2(x + y)$

4. ______________

5. What is the difference of -7 and -15?

5. ______________

6. Evaluate the expression.

$100 - (20)(2) \div 5$

6. ______________

7. Evaluate $(7 + y) \div 10x$ when $x = \frac{2}{7}$ and $y = \frac{1}{7}$.

7. ______________

8. Evaluate the expression to two decimal places.

$\frac{42.18}{4.89 + z}$ when $z = 2.68$

8. ______________

Name ______________________

9. Solve the equation. **9.** ____________

$-3x + 5 = -7x - 4$

10. Solve the equation. **10.** ____________

$5(2 - \frac{1}{2}x) = 4 - (\frac{1}{4}x - 1)$

11. ***Geometry*** Find the area of the shaded region. **11.** ____________

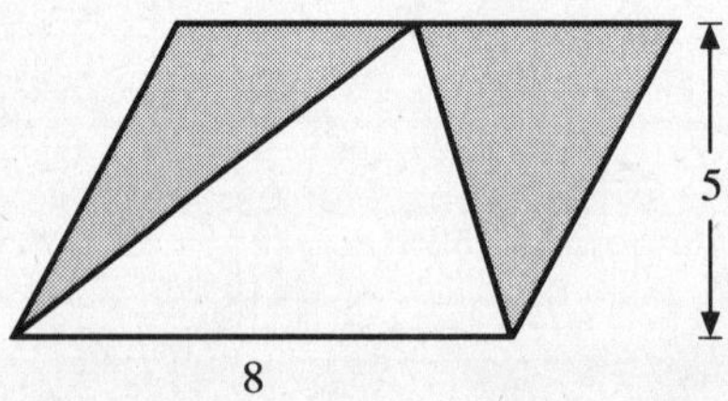

12. Write the following expression using exponents. **12.** ____________

"x is doubled and the result is raised to the nth power."

13. Evaluate $2a^3 + (2a)^3$ when $x = -3$. **13.** ____________

14. ***Population of Hawaii*** For 1980 through 1990, the population, P (in 1000's), of Hawaii can be modeled by **14.** ____________

$P = 17(t + 56.6)$

where $t = 0$ represents 1980. What was the population in 1987?

15. ***Simple Interest Formula*** Solve for P. **15.** ____________

$A = P + Prt$

Name ______________________

16. Solve for z.

$$y^2z + 7z = y$$

16. ______________

17. Solve the inequality and sketch its graph.

$$3(1 + x) > 1 + 5x$$

17. ______________
Use graph at left.

18. Is $x = -7$ a solution of the inequality

$$5x - 7 \leq 3(x - 7)?$$

18. ______________

19. Solve the inequality and sketch its graph.

$$|2x - 5| > 1$$

19. ______________
Use graph at left.

20. ***Stock Investment*** You have \$20,000 available to invest in two stocks, A and B. Write an inequality stating the restriction on A if at least \$3000 must be invested in each stock.

20. ______________

Mid-Chapter 2 Test Form A

Name ____________________

Date ____________________

Use a straight edge to draw straight lines.

1. Plot the points in the coordinate plane. In which quadrant does each point lie?

$(-2,\ 1),\ (3,\ 2),\ (-1,\ -3)$

1. ____________________ *Use graph at left.*

2. Complete the table of values for

$y = -3x + 7.$

x	−2	−1	0	1	2	3
y						

2. *Use the table provided.*

3. Sketch the graph of the equation. Find the slope and the *y*-intercept.

$y = 2x - 1$

3. ____________________ *Use graph at left.*

4. Write the equation of the horizontal line passing through the point $(5,\ -2)$.

4. ____________________

5. Which of the given lines are parallel?

$2x + 4y = 7,\quad 4x - 2y = 7,\quad 4x + 8y = 5$

5. ____________________

6. Write the equation of the line in slope-intercept form.

$5x + 7y = -4$

6. ____________________

7. Write an equation of the line passing through the two points.

$(-4,\ 3),\ (2,\ -1)$

7. ____________________

Mid-Chapter 2 Test Form B

Name ______________________

Date ______________________

Use a straight edge to draw straight lines.

1. Plot the points in the coordinate plane. In which quadrant does each point lie?

$(3,\ 3),\ (2,\ -2),\ (-1,\ -3)$

1. ______________________ *Use graph at left.*

2. Complete the table of values for

$y = 4x - 8.$

x	−2	−1	0	1	2	3
y						

2. *Use the table provided.*

3. Sketch the graph of the equation. Find the slope and the *y*-intercept.

$y = -3x + 3$

3. ______________________ *Use graph at left.*

4. Write the equation of the vertical line passing through the point $(5,\ -2)$.

4. ______________________

5. Which of the given lines are perpendicular?

$2x + 4y = 7, \quad 4x - 2y = 7, \quad 2x - 4y = 7$

5. ______________________

6. Write the equation of the line in slope-intercept form.

$6x - 7y = 4$

6. ______________________

7. Write an equation of the line passing through the two points.

$(-3,\ 2),\ (3,\ 5)$

7. ______________________

Name ______________________

Date ______________________

Use a straight edge to draw straight lines.

1. In which quadrant does the point $(-3, -4)$ lie? **1.** ________

2. Write an equation for the vertical line passing through the point $(-5, 3)$. **2.** ________

3. Sketch the graph of the equation $y = 3x - 1$. Label the intercepts. **3.** *Use graph at left.*

4. Find the slope of the line containing the points. **4.** ________

$(3, -1), (6, 4)$

5. Which line is steeper? **5.** ________

$y = \frac{5}{3}x - 2, \quad y = \frac{3}{2}x + 3$

6. Line 1 contains $(2, 4)$ and $(0, -2)$. Line 2 contains $(-1, -3)$ and $(1, 3)$. Are the lines parallel, perpendicular, or neither? **6.** ________

7. Find the x- and y-intercepts of the line. **7.** ________

$5x - 4y = 20$

8. Write the equation in slope-intercept form. Then identify the slope and y-intercept. **8.** ________

$15x - 3y = 7$

 Name ______________________

9. Sketch the line.

$y = \frac{3}{4}x - 2$

9. *Use graph at left.*

10. The values of x and y vary directly. $y = -6$ when $x = \frac{1}{4}$. Write an equation that relates the variables.

10. ______________________

11. ***Rental Car Charge*** A rental car costs \$25 plus a fixed charge per mile driven. The total charge for 210 miles of use was \$67. Write an equation for the cost, C (in dollars), in terms of the miles driven, x.

11. ______________________

12. Is the ordered pair $(-3, 7)$ a solution of the inequality

$7x - 9y \leq -10$?

12. ______________________

13. Sketch the graph of the inequality.

$\frac{7}{3}x > 7$

13. *Use graph at left.*

14. Sketch the graph of the inequality.

$y \geq \frac{2}{3}x + 2$

14. *Use graph at left.*

15. Find the vertex of the graph.

$y = |3 - x| - 3$

15. ____________________

16. Sketch the graph of the equation.

$y = \frac{1}{2}|x - 2|$

16. *Use graph at left.*

17. ***Population*** The population, P (in 1000s), of a town can be modeled by $P = 2|t - 6| + 4$, where $t = 0$ represents 1990. During which two years does the town have a population of 8000?

17. ____________________

18. For the scatter-plot shown, state whether x and y have a positive correlation, a negative correlation, or no correlation.

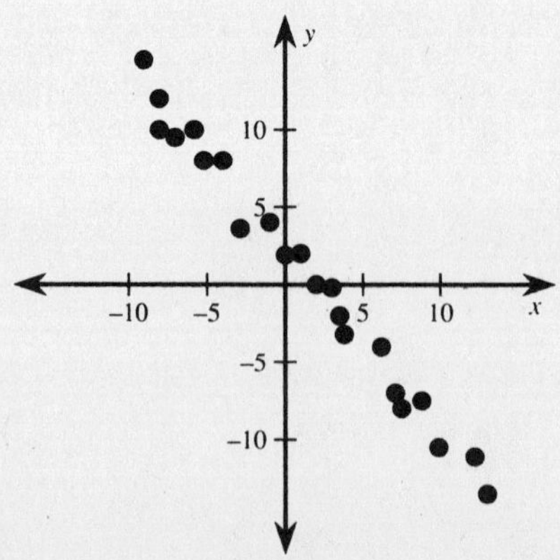

18. ____________________

Chapter 2 Test Form B

(Page 1 of 3 pages)

Name ______________________

Date ______________________

Use a straight edge to draw straight lines.

1. In which quadrant does the point (2, −5) lie? **1.** ______________

2. Write an equation for the horizontal line passing through the point (5, −3). **2.** ______________

3. Sketch the graph of the equation $y = -3x + 2$. Label the intercepts. **3.** *Use graph at left.*

4. Find the slope of the line containing the points. **4.** ______________

(4, −1), (5, −2)

5. Which line is steeper? **5.** ______________

$y = \frac{3}{5}x - 2,\ y = \frac{2}{3}x + 3$

6. Line 1 contains (−3, 5) and (2, 0). Line 2 contains (1, −3) and (−1, 3). Are the lines parallel, perpendicular, or neither? **6.** ______________

7. Find the x- and y-intercepts of the line. **7.** ______________

$3x - 7y = 42$

8. Write the equation in slope-intercept form. Then identify the slope and y-intercept. **8.** ______________

$9x - 5y = 35$

Name ______________________

9. Sketch the line.

$y = \frac{1}{3}x + 1$

9. *Use graph at left.*

10. The values of x and y vary directly. $y = -7$ when $x = 14$. Write an equation that relates the variables.

10. ______________________

11. ***Rental Car Charge*** A rental car costs \$25 plus a fixed charge per mile driven. The total charge for 320 miles of use was \$89. Write an equation for the cost, C (in dollars), in terms of the miles driven, x.

11. ______________________

12. Is the ordered pair $(2, -1)$ a solution of the inequality

$5x - 3y \leq 10$?

12. ______________________

13. Sketch the graph of the inequality.

$\frac{7}{3}x \leq 7$

13. *Use graph at left.*

Name ______________________

14. Sketch the graph of the inequality.

$y \leq \frac{2}{3}x - 2$

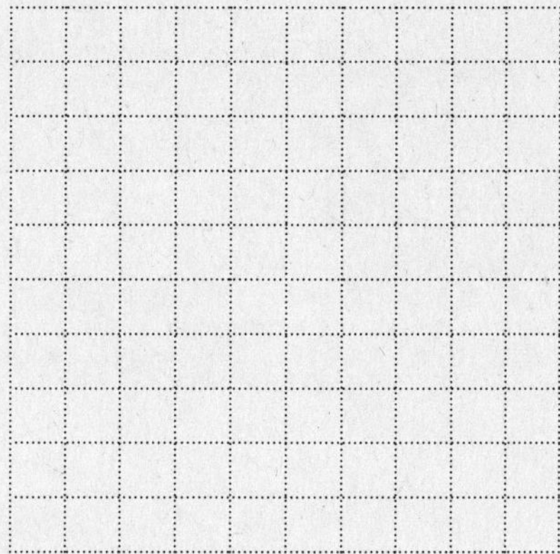

14. *Use graph at left.*

15. Find the vertex of the graph.

$y = -|x| + 4$

15. ______________________

16. Sketch the graph of the equation.

$y = 3 - |x|$

16. *Use graph at left.*

17. ***Population*** The population, P (in 1000s), of a town can be modeled by $P = 2|t - 8| + 4$, where $t = 0$ represents 1990. During which two years does the town have a population of 8000?

17. ______________________

18. For the scatter-plot shown, state whether x and y have a positive correlation, a negative correlation, or no correlation.

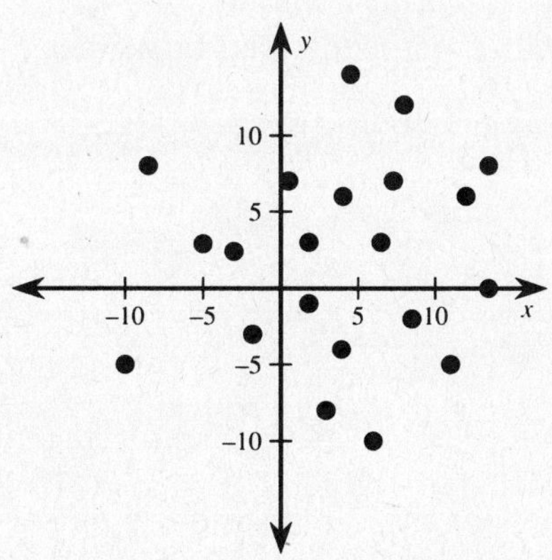

18. ______________________

Chapter 2 Test

Form C

Name ____________________

Date ____________________

Use a straight edge to draw straight lines.

1. In which quadrant does the point $(-3,\ 0)$ lie? **1.** ____________

2. Write an equation for the y-axis. **2.** ____________

3. Sketch the graph of the equation $y = -2x + 5$. Label the intercepts. **3.** *Use graph at left.*

4. Find the slope of the line containing the points. **4.** ____________

   $(-11,\ 7),\ (-6,\ 10)$

5. Which line is steeper? **5.** ____________

   $y = \frac{7}{4}x + 5, \quad y = \frac{5}{3}x - 2$

6. Line 1 contains $(2,\ -4)$ and $(0,\ 2)$. Line 2 contains $(-4,\ 5)$ and $(-1,\ 6)$. Are the lines parallel, perpendicular, or neither? **6.** ____________

7. Find the x- and y-intercepts of the line. **7.** ____________

   $4x - 3y = 36$

8. Write the equation in slope-intercept form. Then identify the slope and y-intercept. **8.** ____________

   $18x - 42y = 24$

Name ______________________

9. Sketch the line.

$y = -\frac{2}{3}x - 2$

9. *Use graph at left.*

10. The values of x and y vary directly. $y = -8$ when $x = 5$. Write an equation that relates the variables.

10. ______________________

11. ***Car Rental Cost*** A rental car costs a minimum charge plus a fixed charge per mile driven. One customer was charged \$60 for 160 miles of use. Another customer drove 280 miles and paid \$90. Write an equation relating the total cost, C (in dollars), to the miles driven, x.

11. ______________________

12. Is the ordered pair $(-7,\ 2)$ a solution of the inequality

$3x + 10y \geq -1$

12. ______________________

13. Sketch the graph of the inequality.

$\frac{2}{3}y > -\frac{4}{3}$

13. *Use graph at left.*

Name ______________________

14. Sketch the graph of the inequality.

$3x + 2y < -6$

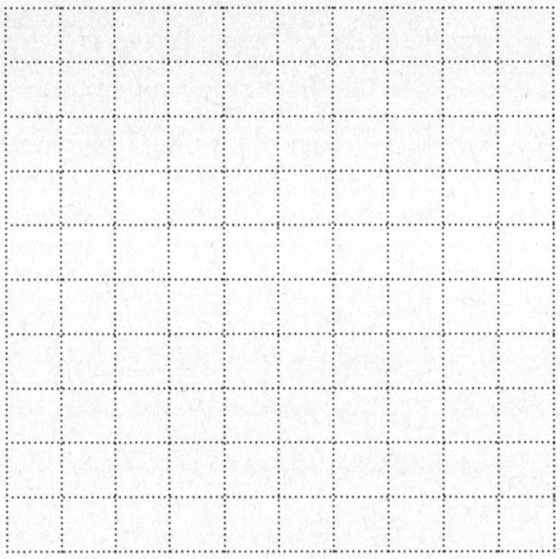

14. *Use graph at left.*

15. Find the vertex of the graph.

$y = -2|3x - 1| + 2$

15. ______________________

16. Sketch the graph of the equation.

$y = |x + 2| - 2$

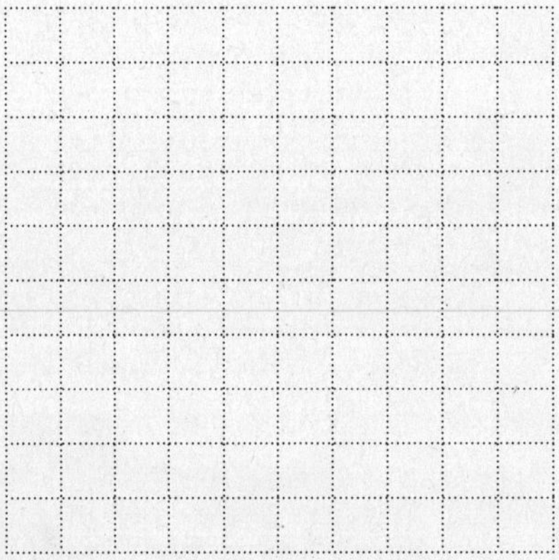

16. *Use graph at left.*

17. ***Population*** The population, P (in 1000s), of a town can be modeled by $P = 2|t - 5| + 6$, where $t = 0$ represents 1990. During which two years does the town have a population of 12,000?

17. ______________________

18. For the scatter-plot shown, state whether x and y have a positive correlation, a negative correlation, or no correlation.

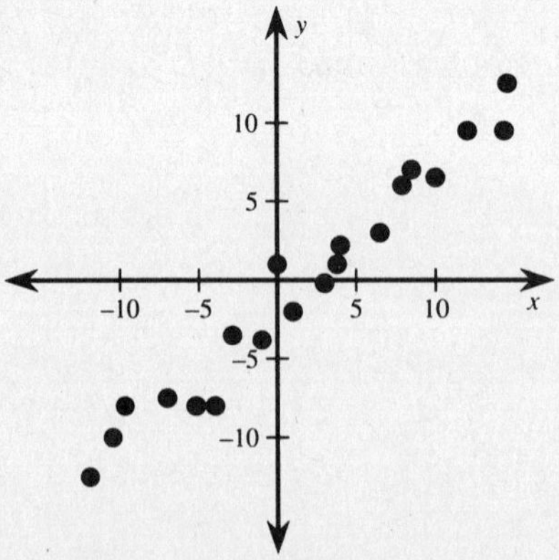

18. ______________________

Mid-Chapter 3 Test Form A

Name ______________________

Date ______________________

Use a straight edge to draw straight lines.

1. How many solutions does the linear system have?

$$\begin{cases} 3x - 4y = 3 \\ 3x + 4y = 3 \end{cases}$$

1. ______________________

2. Is $(-3,\ 4)$ a solution of the system?

$$\begin{cases} 5x + 4y = 1 \\ -3x + 2y = -1 \end{cases}$$

2. ______________________

3. Sketch the graph of the system. Estimate the solution.

$$\begin{cases} x + 2y = 2 \\ x + y = 3 \end{cases}$$

3. ______________________
Use graph at left.

4. Use substitution to solve the system.

$$\begin{cases} 3x - y = 4 \\ 2x + 3y = 10 \end{cases}$$

4. ______________________

5. Solve the system using a linear combination.

$$\begin{cases} 6x + 3y = -1 \\ 3x - 6y = 12 \end{cases}$$

5. ______________________

6. ***VCR Sales*** A dealer carries two models of VCRs. One sells for \$300 and the other sells for \$225. The dealer sells 22 VCRs for a total revenue of \$5700. How many of each type did the dealer sell?

6. ______________________

Mid-Chapter 3 Test Form B

Name ______________________

Date ______________________

Use a straight edge to draw straight lines.

1. How many solutions does the linear system have?

$$\begin{cases} 3x - 4y = 3 \\ 3x - 4y = -3 \end{cases}$$

1. ______________

2. Is $(-3, 4)$ a solution of the system?

$$\begin{cases} 5x + 4y = 1 \\ -3x - 2y = -1 \end{cases}$$

2. ______________

3. Sketch the graph of the system. Estimate the solution.

$$\begin{cases} x - 2y = -2 \\ x - y = -3 \end{cases}$$

3. ______________
Use graph at left.

4. Use substitution to solve the system.

$$\begin{cases} 5x + y = 14 \\ 2x + 3y = 3 \end{cases}$$

4. ______________

5. Solve the system using a linear combination.

$$\begin{cases} 3x - 2y = -2 \\ 9x + 4y = 9 \end{cases}$$

5. ______________

6. ***VCR Sales*** A dealer carries two models of VCRs. One sells for \$300 and the other sells for \$225. The dealer sells 24 VCRs for a total revenue of \$6000. How many of each type did the dealer sell?

6. ______________

Chapter 3 Test Form A

(Page 1 of 3 pages)

Name ______________________

Date ______________________

Use a straight edge to draw straight lines.

1. Are there none, one, or many solutions to the system?

$$\begin{cases} x - 4y = 2 \\ 2x - 8y = 5 \end{cases}$$

1. ______________________

2. Is $(5, -2)$ a solution of the system?

$$\begin{cases} 2x + 6y = -2 \\ x + 2y = 1 \end{cases}$$

2. ______________________

3. Sketch the graph of the system. Estimate the solution.

$$\begin{cases} 2x - 3y + 6 = 0 \\ 5x - 2y - 7 = 0 \end{cases}$$

3. ______________________
Use graph at left.

4. ***Theater Tickets*** 1500 theater tickets were sold for a performance. General admission was \$12 but student rates offered a 50% discount. Box office receipts totaled \$16,200. How many students attended?

4. ______________________

5. Solve the system.

$$\begin{cases} y = -4x + 4 \\ y = -x - 5 \end{cases}$$

5. ______________________

 Name ______________________

6. Solve the linear system.

$$\begin{cases} 4x - 3y = -1 \\ 3x + 4y = -3 \end{cases}$$

6. ______________________

7. ***Geometry*** The measures of the two acute angles of a right triangle differ by 19°. What are their measures?

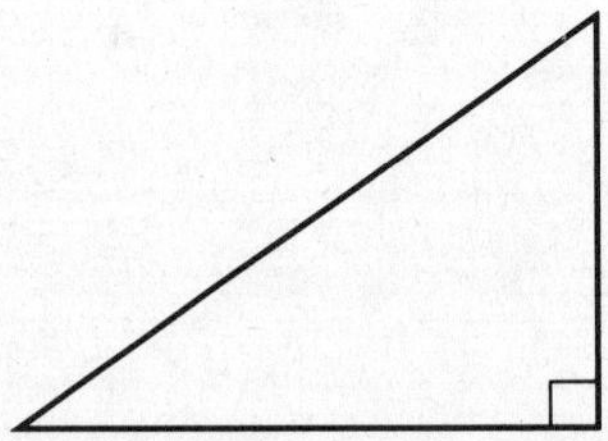

7. ______________________

8. Sketch the graph of the system of linear inequalities.

$$\begin{cases} x > -3 \\ y \geq \ \ 1 \end{cases}$$

8. *Use graph at left.*

9. Sketch the graph of the system described.

"x and y are each greater than -3 but not greater than 2."

9. *Use graph at left.*

 Name ____________

10. Sketch the graph of the system of linear inequalities.

$$\begin{cases} y \le \frac{1}{2}x + 2 \\ y \ge -\frac{1}{2}x - 2 \\ x \le 3 \end{cases}$$

Label the vertices.

10. *Use graph at left.*

11. Find the *maximum* value of C under the constraints.

$C = -2x + 3y$ Constraints:

$$\begin{cases} x \ge -4 \\ x \le 3 \\ y \ge -1 \\ y \le 2 \end{cases}$$

11. ____________
Use graph at left.

12. Solve the linear system.

$$\begin{cases} x + 2y - 4z = -12 \\ -x + z = 1 \\ x + y + z = 4 \end{cases}$$

12. ____________

Chapter 3 Test

Form B

(Page 1 of 3 pages)

Name ____________________

Date ____________________

Use a straight edge to draw straight lines.

1. Are there none, one, or many solutions to the system?

$$\begin{cases} x - 4y = 2 \\ x + 4y = 2 \end{cases}$$

1. ____________________

2. Is $(5, \ -2)$ a solution of the system?

$$\begin{cases} 2x + 6y = -2 \\ 2x + y = 6 \end{cases}$$

2. ____________________

3. Sketch the graph of the system. Estimate the solution.

$$\begin{cases} 3x - 2y = -7 \\ x + y = 1 \end{cases}$$

3. ____________________
Use graph at left.

4. ***Theater Tickets*** 1500 theater tickets were sold for a performance. General admission was \$12 but student rates offered a 50% discount. Box office receipts totaled \$15,600. How many students attended?

4. ____________________

5. Solve the system.

$$\begin{cases} 3x + 4y = -3 \\ 2x + y = 8 \end{cases}$$

5. ____________________

Name ______________________

6. Solve the linear system.

$$\begin{cases} 2x + 5y = -2 \\ 3x - 2y = 4 \end{cases}$$

6. ______________________

7. ***Geometry*** The measures of the two acute angles of a right triangle differ by 15°. What are their measures?

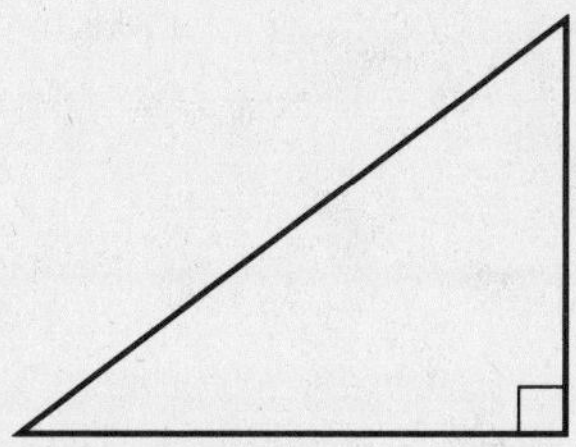

7. ______________________

8. Sketch the graph of the system of linear inequalities.

$$\begin{cases} x \geq -2 \\ y < 4 \end{cases}$$

8. *Use graph at left.*

9. Sketch the graph of the system described.

"x and y are each less than 4 and not negative."

9. *Use graph at left.*

Name ____________________

10. Sketch the graph of the system of linear inequalities.

$$\begin{cases} y \geq \frac{1}{2}x + 2 \\ y \geq -\frac{1}{2}x + 2 \\ y \leq 4 \end{cases}$$

Label the vertices.

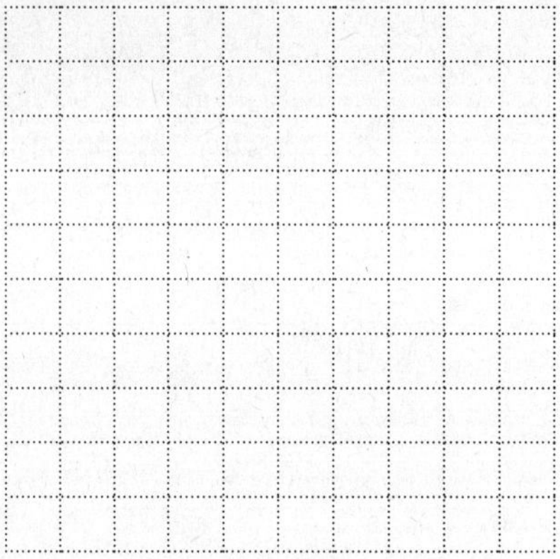

10. *Use graph at left.*

11. Find the *minimum* value of C under the constraints.

$C = -2x + 3y$ Constraints:

$$\begin{cases} x \geq -4 \\ x \leq 3 \\ y \geq -1 \\ y \leq 2 \end{cases}$$

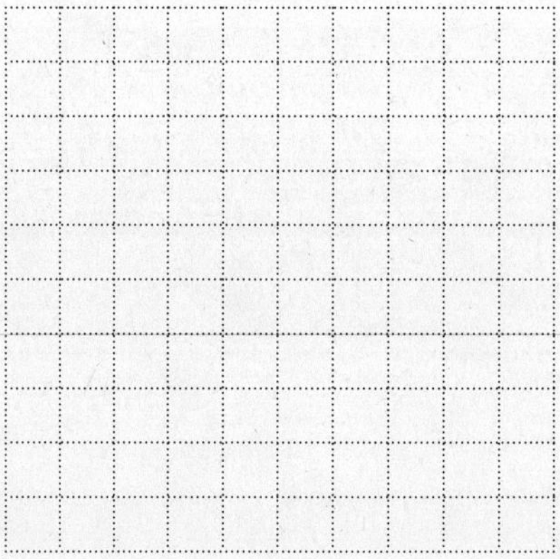

11. ____________________
Use graph at left.

12. Solve the linear system.

$$\begin{cases} 2x + y - z = -1 \\ x - 2y = -7 \\ x + y + z = 4 \end{cases}$$

12. ____________________

Chapter 3 Test

Form C

(Page 1 of 3 pages)

Name ______________________

Date ______________________

Use a straight edge to draw straight lines.

1. Are there none, one, or many solutions to the system?

$$\begin{cases} \frac{1}{2}x - 2y = \frac{1}{3} \\ 3x - 12y = 2 \end{cases}$$

1. ______________________

2. Is $(2, -1)$ a solution of the system?

$$\begin{cases} 2x + 6y = -2 \\ x + 2y = 1 \end{cases}$$

2. ______________________

3. Sketch the graph of the system. Estimate the solution.

$$\begin{cases} 2x - 3y + 5 = 0 \\ 5x - 2y - 4 = 0 \end{cases}$$

3. ______________________
Use graph at left.

4. ***Theater Tickets*** 1500 theater tickets were sold for a performance. General admission was \$12 but student rates offered a 50% discount. Box office receipts totaled \$15,480. How many students attended?

4. ______________________

5. Solve the system.

$$\begin{cases} y = -\frac{3}{4}x + \frac{1}{4} \\ y = \frac{3}{4}x - \frac{3}{4} \end{cases}$$

5. ______________________

Name ______________________

6. Solve the linear system.

$$\begin{cases} 7x - 4y = 12 \\ 3x + 2y = 4 \end{cases}$$

6. ______________

7. ***Geometry*** The measures of the two acute angles of a right triangle differ by 43°. What are their measures?

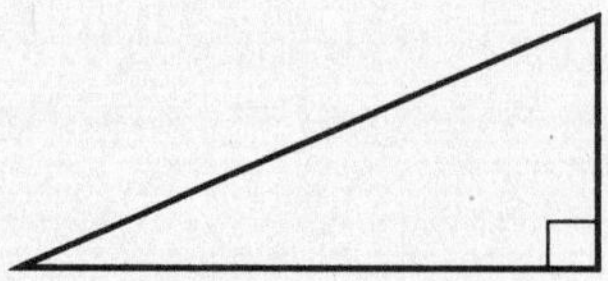

7. ______________

8. Sketch the graph of the system of linear inequalities.

$$\begin{cases} 2x + 1 \geq 0 \\ 3y - 6 > 0 \end{cases}$$

8. *Use graph at left.*

9. Sketch the graph of the system described.

"x and y are each less than 3 but no less than -2."

9. *Use graph at left.*

10. Sketch the graph of the system of linear inequalities.

$$\begin{cases} y \le -x + 4 \\ y \le 2x + 4 \\ y \ge -2 \end{cases}$$

Label the vertices.

10. *Use graph at left.*

11. Find the *maximum* and *minimum* values of C under the constraints.

$C = -2x + 3y$ Constraints:

$$\begin{cases} x \ge -2 \\ x \le 4 \\ y \ge -2 \\ y \le 2 \end{cases}$$

11. ______________________
Use graph at left.

12. Solve the linear system.

$$\begin{cases} 2x + 3y - z = 1 \\ x + y + z = 3 \\ 3x - y + z = 15 \end{cases}$$

12. ______________________

Cumulative Test

Chapters 1-3

Name ______________________

Date ______________________

1. Plot the numbers on a real number line. Then decide which is the greatest.

   $\frac{3}{2}, -\frac{5}{3}, \frac{3}{5}$

1. ______________ *Use graph at left.*

2. Write the numbers in *decreasing* order.

   $-\frac{7}{4}, \frac{3}{5}, 0, -\frac{1}{5}, \frac{5}{8}, -\frac{7}{5}, \frac{5}{7}$

2. ______________

3. State the property that is illustrated.

   $(7 \cdot 4) \cdot 8 = 7 \cdot (4 \cdot 8)$

3. ______________

4. What is the difference of -3 and 7?

4. ______________

5. Evaluate $96 - (80)(6) \div 5$.

5. ______________

6. Evaluate $(8 - 3x) \div 2y$ when $x = -8$ and $y = \frac{2}{3}$.

6. ______________

 Name ______________________

7. Evaluate the expression to two decimal places. **7.** ______________

$z \cdot (85.13) - (7.14)(2.31)$ when $z = 1.27$.

8. Solve the equation. **8.** ______________

$9x - 3 = 7x + 5$

9. Solve the equation. **9.** ______________

$7(2 - 3x) = 6 - (5 - 2x)$

10. Write the expression using exponents. **10.** ______________

"5 cubed, times y to the nth power"

11. Solve for r. **11.** ______________

$A = P(1 + rt)$

12. ***County Population*** For 1980 through 1990, the population of a county can be modeled by $P = 132 + 3t$ where $t = 0$ corresponds to 1980 and P is in thousands. Approximate the population of the county in 1987. **12.** ______________

 Name ____________

13. Solve and graph the inequality
$x - 5 > 2(2 - 4x)$.

13. ____________ *Use graph at left.*

14. Is $-\frac{5}{2}$ a solution of the inequality
$2x - 3 \leq 1 - 5x$?

14. ____________

15. Solve and sketch the graph of the inequality
$|2x - 5| \leq 3$.

15. ____________ *Use graph at left.*

16. Complete the table of values for
$y = -\frac{3}{2}x + \frac{7}{2}$.

x	−2	−1	0	1	2	3
y						

16. *Use table at left.*

17. Sketch the graph of the equation. Find the slope and the y-intercept.
$y = \frac{1}{2}x - 1$

17. ____________ *Use graph at left.*

18. Write the equation of the vertical line passing through the point $(7, -3)$.

18. ____________

19. Write the equation of the line

$$6x + 5y = 20$$

in slope-intercept form.

19. ______

20. Which of the given lines are perpendicular?

$$2x + 4y = 5, \quad 4x - 2y = 3, \quad 2x - 4y = 7$$

20. ______

21. Write an equation of the line passing through the points

$(-1, 7)$ and $(1, 1)$.

21. ______

22. ***Car Travel*** Two cars are traveling the same direction on a straight highway. At 1 P.M. car A is 50 miles behind car B. At 2:30 P.M. car A is 17 miles behind car B. Write an equation for the distance, D (in miles), separating the two cars in terms of time t (assume $t = 0$ at 1 P.M.).

22. ______

23. Find the x- and y-intercepts of the line

$$10x - 2y = 15.$$

23. ______

 Name ______________________

24. The values of x and y vary *directly*.

$y = -9$ when $x = \frac{1}{5}$

Write an equation that relates the variables.

24. ______________________

25. ***Telephone Cost*** A telephone call costs \$2.50 for the first 3 minutes and a fixed charge for each additional minute. A 10-minute call costs \$6.70. Write an equation for the cost, C (in dollars), in terms of the time t (in minutes for $t \geq 3$).

25. ______________________

26. Is the ordered pair $(-5, -6)$ a solution of the inequality

$-3x + 5y \geq -10$?

26. ______________________

27. Sketch the graph of the inequality.

$\frac{5}{8}x < \frac{15}{4}$

27. *Use graph at left.*

Name ______________________

28. Sketch the graph of the inequality

$y \geq \frac{1}{2}x - 3.$

28. *Use graph at left.*

29. Evaluate $2|1 - 2x| - 3$ when $x = 4$.

29. ______________________

30. Find the vertex of the graph.

$y = 5 - |1 - 2x|$

30. ______________________

31. Sketch the graph of the equation.

$y = 2|x + 1|$

31. ______________________ *Use graph at left.*

32. For the scatter-plot shown, state whether x and y have a positive correlation, a negative correlation, or no correlation.

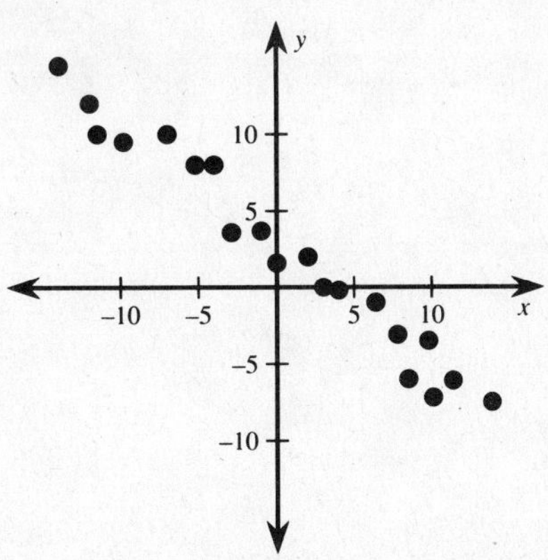

32. ______________________

 Name ______________________

33. Are there none, one, or many solutions to the system?

$$\begin{cases} 2x - 3y = 5 \\ 4x - 6y = 5 \end{cases}$$

33. ______________________

34. Sketch the graph of the system.
Estimate the solution.

$$\begin{cases} 2x + 3y = 3 \\ 2x - 3y = 9 \end{cases}$$

34. ______________________
Use graph at left.

35. ***Number of Coins*** A mixture of dimes and quarters has a total value of $10.20. There are 57 coins in all. How many of each type of coin is present?

35. ______________________

36. Solve the system.

$$\begin{cases} 3x + 4y = 7 \\ 2x - 5y = 20 \end{cases}$$

36. ______________________

Name ____________________

37. ***Geometry*** The measure of the two acute angles of a right triangle differ by $31°$. What are their measures?

37. ____________________

38. Sketch the graph of the system described.

$$\begin{cases} x > -3 \\ y \geq -1 \end{cases}$$

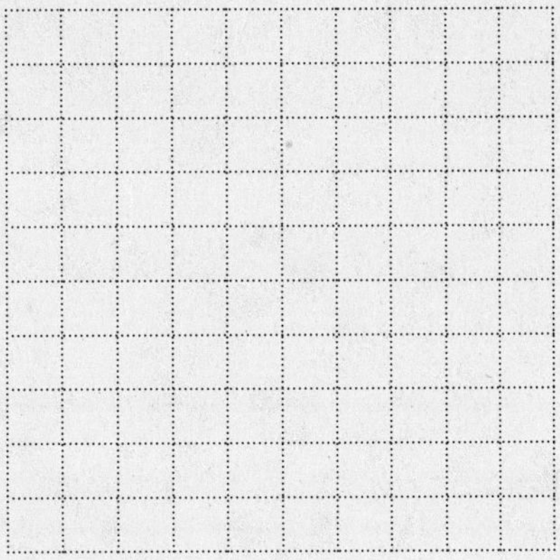

38. *Use graph at left.*

39. Sketch the graph of the system of linear inequalities.

$$\begin{cases} y \geq 2x - 4 \\ y \geq -2x - 4 \\ y \leq 2 \end{cases}$$

Label the vertices.

39. *Use graph at left.*

40. Find the *maximum* value of C under the constraints.

$C = -5x + 7y$ Constraints

$$\begin{cases} y \leq x - 1 \\ x \leq 4 \\ y \geq -3 \end{cases}$$

40. ____________________
Use graph at left.

Mid-Chapter 4 Test Form A

Name ______________________

Date ______________________

1. State the order of the matrix. $\begin{bmatrix} 2 & 1 & -7 \\ 1 & 4 & 3 \end{bmatrix}$ **1.** ______________

In 2–5, perform the indicated matrix operations, if possible.

2. $\begin{bmatrix} 2 & -1 & 4 \\ -3 & 2 & -2 \end{bmatrix} + \begin{bmatrix} 3 & 1 & -5 \\ 2 & 4 & 7 \end{bmatrix} - \begin{bmatrix} 1 & -1 & 2 \\ 3 & 0 & -1 \end{bmatrix}$ **2.** ______________

3. $\begin{bmatrix} 3 & 1 \\ 2 & -2 \end{bmatrix}\begin{bmatrix} 1 & -1 \\ 2 & 0 \end{bmatrix}$ **3.** ______________

4. $3\begin{bmatrix} -2 & -1 & 5 \\ -1 & 2 & 0 \end{bmatrix}$ **4.** ______________

5. $\begin{bmatrix} 2 & 1 & -4 \end{bmatrix}\begin{bmatrix} 2 \\ 5 \end{bmatrix}$ **5.** ______________

6. Evaluate $\begin{vmatrix} 2 & 3 \\ -4 & 1 \end{vmatrix}$. **6.** ______________

7. Evaluate the determinant of the matrix. $\begin{bmatrix} 2 & -1 & 5 \\ 2 & 0 & 4 \\ 3 & -5 & -2 \end{bmatrix}$ **7.** ______________

8. Find the inverse of $\begin{bmatrix} 2 & 3 \\ -1 & 4 \end{bmatrix}$. **8.** ______________

9. Solve the equation. $\begin{bmatrix} 2 & 3 \\ -1 & 4 \end{bmatrix}\begin{bmatrix} x \\ y \end{bmatrix} = \begin{bmatrix} 7 \\ 2 \end{bmatrix}$ **9.** ______________

10. ***Geometry*** Find the area of the triangle. **10.** ______________

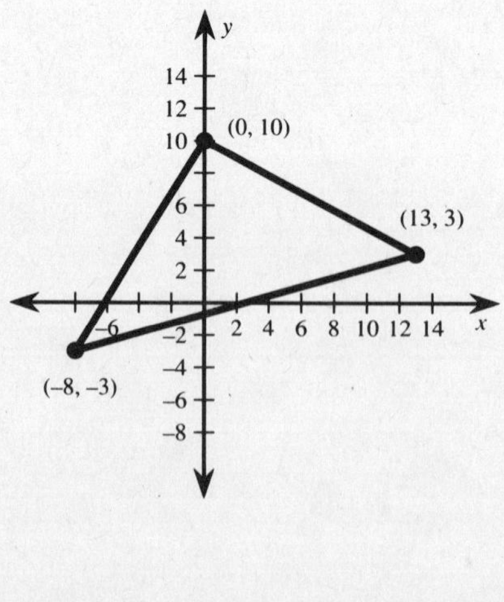

Mid-Chapter 4 Test Form B

Name ______________________

Date ______________________

1. State the order of the matrix. $\begin{bmatrix} 2 & 1 & 7 & 9 \\ 8 & 0 & 4 & -3 \end{bmatrix}$ 1. ____________

In 2–5, perform the indicated matrix operations, if possible.

2. $\begin{bmatrix} 3 & -1 & 5 \\ 2 & 1 & 0 \end{bmatrix} - \begin{bmatrix} 7 & 1 & 4 \\ 0 & 3 & 5 \end{bmatrix} + \begin{bmatrix} 1 & 2 & -3 \\ 4 & 0 & 2 \end{bmatrix}$ 2. ____________

3. $\begin{bmatrix} 2 & -1 \\ 0 & 3 \end{bmatrix}\begin{bmatrix} 1 & 2 \\ -1 & 4 \end{bmatrix}$ 3. ____________

4. $\begin{bmatrix} 3 & -1 & 7 \end{bmatrix}\begin{bmatrix} 3 \\ 5 \end{bmatrix}$ 4. ____________

5. $-4\begin{bmatrix} 1 & 1 & -2 \\ 2 & 0 & -1 \end{bmatrix}$ 5. ____________

6. Evaluate $\begin{vmatrix} 2 & -6 \\ -1 & 4 \end{vmatrix}$. 6. ____________

7. Evaluate the determinant of the matrix. $\begin{bmatrix} 2 & 1 & -1 \\ 1 & 4 & 2 \\ -2 & 0 & 3 \end{bmatrix}$ 7. ____________

8. Find the inverse of $\begin{bmatrix} 7 & 4 \\ -3 & 1 \end{bmatrix}$. 8. ____________

9. Solve for x and y. $\begin{bmatrix} 7 & 4 \\ -3 & 1 \end{bmatrix}\begin{bmatrix} x \\ y \end{bmatrix} = \begin{bmatrix} 3 \\ -4 \end{bmatrix}$ 9. ____________

10. ***Geometry*** Find the area of the triangle. 10. ____________

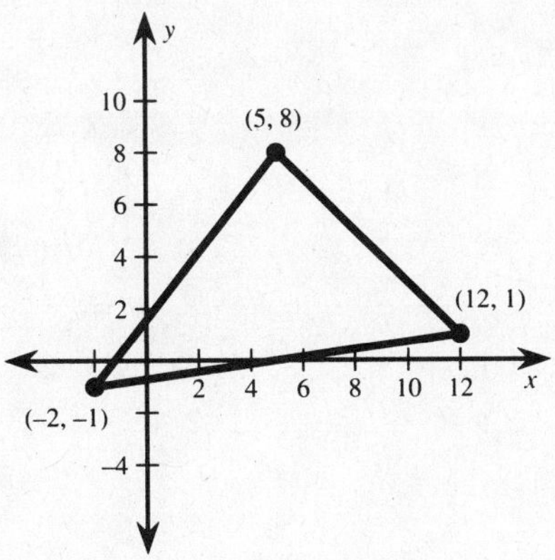

Chapter 4 Test Form A

(Page 1 of 4 pages)

Name ______________________

Date ______________________

1. State the order of the matrix. $\begin{bmatrix} 2 & 3 \\ 1 & 7 \\ -4 & 6 \end{bmatrix}$ **1.** ______________

In 2–6, perform the matrix operations, if possible.

2. $\begin{bmatrix} 3 & 1 & -5 \\ 7 & 2 & 4 \\ 1 & 8 & 3 \end{bmatrix} + \begin{bmatrix} 1 & 1 & -2 \\ 3 & 7 & 0 \\ 2 & 4 & -1 \end{bmatrix}$ **2.** ______________

3. $\begin{bmatrix} 4 & -7 \\ 10 & 5 \end{bmatrix} - \begin{bmatrix} 8 & 10 \\ 1 & 9 \end{bmatrix}$ **3.** ______________

4. $5\begin{bmatrix} 2 & 1 & -5 \\ 6 & -3 & 4 \end{bmatrix}$ **4.** ______________

5. $\begin{bmatrix} 2 & -3 \\ 6 & 5 \\ 1 & 4 \end{bmatrix}\begin{bmatrix} 2 & 4 \\ 3 & -2 \end{bmatrix}$ **5.** ______________

6. $\begin{bmatrix} 2 & 5 \\ -1 & 7 \end{bmatrix}\begin{bmatrix} 2 & 1 & -4 \end{bmatrix}$ **6.** ______________

7. Solve for x and y. $2\begin{bmatrix} 1 & 5 \\ 3 & -2 \end{bmatrix} - \begin{bmatrix} -2 & 1 \\ 1 & -3 \end{bmatrix} = \begin{bmatrix} x & 9 \\ 5 & y \end{bmatrix}$ **7.** ______________

8. Evaluate $\begin{vmatrix} 5 & 8 \\ -2 & 3 \end{vmatrix}$. **8.** ______________

9. Evaluate the determinant of the matrix. $\begin{bmatrix} 2 & 0 & -5 \\ 3 & 3 & 1 \\ -2 & 4 & 1 \end{bmatrix}$ **9.** ____________

10. Find the inverse of $\begin{bmatrix} 3 & -2 \\ 7 & 1 \end{bmatrix}$. **10.** ____________

11. ***Geometry*** Find the area of the triangle. **11.** ____________

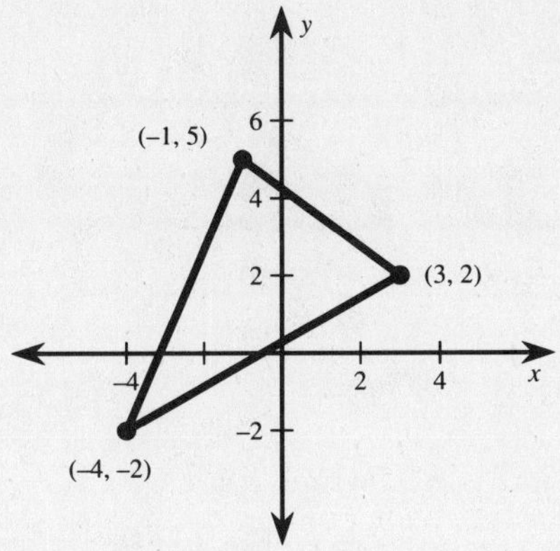

12. ***Inventory*** A company stocks items A, B, and C at each of its two stores. Use matrix multiplication to determine the value of the inventory at each store. **12.** ____________

Inventory

	Item *A*	**Item *B***	**Item *C***
Store *1*	7	9	8
Store *2*	10	5	9

Value of Each Item ($)

Item *A*	10
Item *B*	8
Item *C*	7

Inventory Value ($)

Store *1*	
Store *2*	

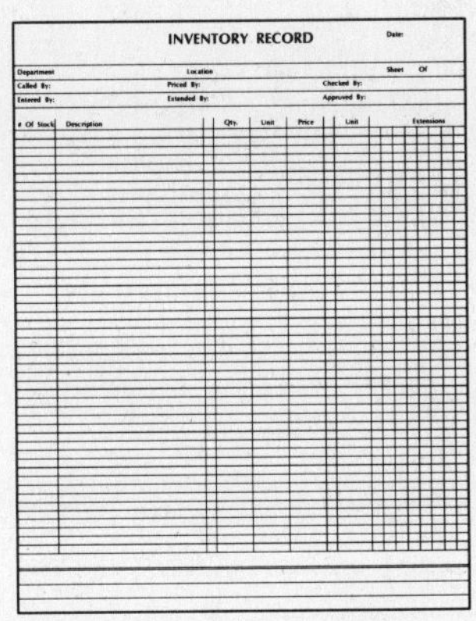

13. Write the linear system as a matrix equation. **13.** ____________

$$\begin{cases} 5x - 2y = 8 \\ 4x + 3y = 7 \end{cases}$$

14. Use an inverse matrix to solve the linear system. **14.** ____________

$$\begin{cases} 7x - 8y = -21 \\ x - y = -2 \end{cases}$$

15. Write the augmented matrix for the linear system. **15.** ____________

$$\begin{cases} x - 2y + z = 4 \\ 2x + y = 8 \\ x - 3y + 2z = -1 \end{cases}$$

16. Write the linear system which is represented by the augmented matrix. **16.** ____________

$$\left[\begin{array}{ccc:c} 1 & 5 & -7 & 3 \\ 0 & 1 & 3 & -4 \\ 0 & 0 & 1 & 7 \end{array}\right]$$

17. Use an augmented matrix to solve the system. **17.** ____________

$$\begin{cases} 2x - y - z = 5 \\ x + y + z = 7 \\ 3x - 2y - 3z = 1 \end{cases}$$

Name ______________________

18. Find the coefficient matrix and evaluate its determinant.

$$\begin{cases} 3x - 5y = 7 \\ 2x + 4y = -6 \end{cases}$$

18. ______________________

19. Use Cramer's Rule to solve the linear system.

$$\begin{cases} 5x + 4y = 0 \\ 6x + 5y = -2 \end{cases}$$

19. ______________________

20. Use Cramer's Rule to solve the linear system.

$$\begin{cases} x + y + z = -4 \\ 2x - y + 3z = 1 \\ 3x + 2y - z = 4 \end{cases}$$

20. ______________________

Chapter 4 Test

Form B

(Page 1 of 4 pages)

Name ______________________

Date ______________________

1. State the order of the matrix. $\begin{bmatrix} 2 & 1 & 4 & 0 \\ 1 & 1 & -2 & 2 \\ 3 & 1 & 4 & 3 \end{bmatrix}$ **1.** __________

In 2–6, perform the matrix operations, if possible.

2. $\begin{bmatrix} 5 & -1 & 2 \\ 4 & 0 & 7 \\ 2 & -1 & 5 \end{bmatrix} + \begin{bmatrix} 5 & 0 & 9 \\ -2 & 4 & 10 \\ 4 & -2 & 3 \end{bmatrix}$ **2.** __________

3. $\begin{bmatrix} -2 & 7 \\ 3 & -5 \end{bmatrix} - \begin{bmatrix} 2 & 4 \\ -1 & 3 \end{bmatrix}$ **3.** __________

4. $-3\begin{bmatrix} 2 & 1 & -4 \\ 1 & 0 & 2 \end{bmatrix}$ **4.** __________

5. $[2 \quad -1 \quad 2]\begin{bmatrix} -2 \\ 4 \\ 6 \end{bmatrix}$ **5.** __________

6. $\begin{bmatrix} 3 & 7 \\ 1 & 5 \end{bmatrix}[2 \quad 9 \quad -1]$ **6.** __________

7. Solve for x and y. $3\begin{bmatrix} 4 & x \\ -1 & 3 \end{bmatrix} - 2\begin{bmatrix} 1 & 3 \\ y & 0 \end{bmatrix} = \begin{bmatrix} 10 & 6 \\ -7 & 9 \end{bmatrix}$ **7.** __________

8. Evaluate $\begin{vmatrix} 6 & -7 \\ 5 & 8 \end{vmatrix}$. **8.** __________

 Name ______________________

9. Evaluate the determinant of the matrix. $\begin{bmatrix} 2 & -1 & 3 \\ 1 & 2 & 2 \\ -2 & 1 & 0 \end{bmatrix}$ **9.** ______________

10. Find the inverse of $\begin{bmatrix} 5 & 7 \\ 9 & 3 \end{bmatrix}$. **10.** ______________

11. ***Geometry*** Find the area of the triangle. **11.** ______________

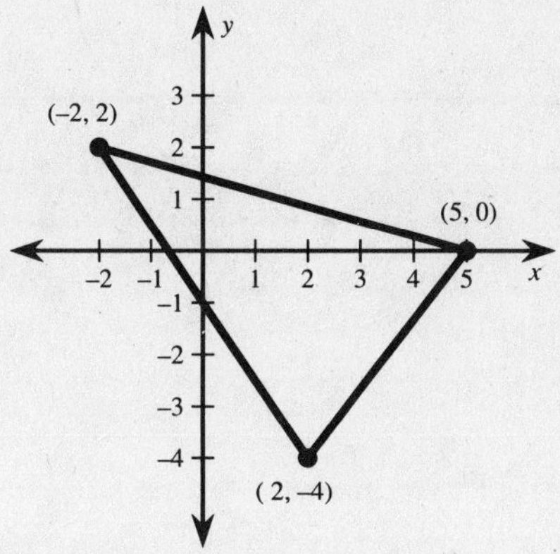

12. ***Inventory*** A company stocks items A, B, and C at each of its two stores. Use matrix multiplication to determine the value of the inventory at each store. **12.** ______________

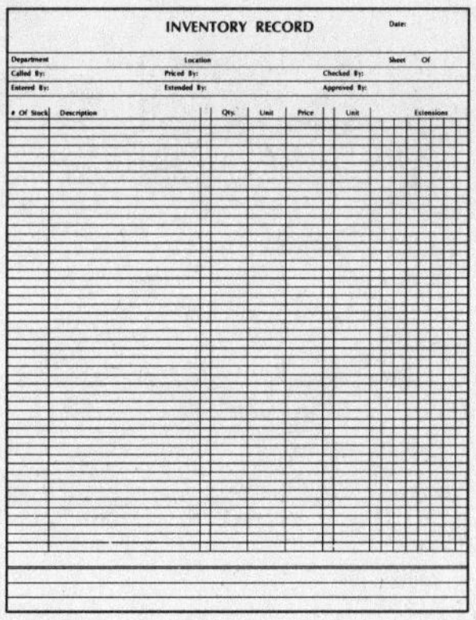

Inventory

	Item *A*	Item *B*	Item *C*
Store 1	7	5	10
Store 2	4	9	5

Value of Each Item ($)

Item A	8
Item B	10
Item C	12

Inventory Value ($)

Store 1	
Store 2	

 Name ____________________

13. Write the linear system as a matrix equation. **13.** ____________

$$\begin{cases} 4x - 10y = 7 \\ 3x + 9y = 4 \end{cases}$$

14. Use an inverse matrix to solve the linear system. **14.** ____________

$$\begin{cases} 3x + y = 3 \\ 5x + 2y = 2 \end{cases}$$

15. Write the augmented matrix for the linear system. **15.** ____________

$$\begin{cases} x - 3y + 4z = 2 \\ 2x + y - z = -1 \\ x + 2y + z = 3 \end{cases}$$

16. Write the linear system which is represented by the augmented matrix. **16.** ____________

$$\left[\begin{array}{ccc|c} 1 & -3 & 2 & 7 \\ 0 & 1 & 5 & -2 \\ 0 & 0 & 1 & 8 \end{array}\right]$$

17. Use the augmented matrix to solve the system. **17.** ____________

$$\begin{cases} x + y + z = 7 \\ 2x - 3y - 2z = 0 \\ 2x - 2y - z = 4 \end{cases}$$

Name ____________________

18. Find the coefficient matrix and evaluate its determinant.

$$\begin{cases} 8x - 5y = 7 \\ 12x + 7y = -4 \end{cases}$$

18. ____________________

19. Use Cramer's Rule to solve the linear system.

$$\begin{cases} 2x + y = 1 \\ 5x + 2y = 5 \end{cases}$$

19. ____________________

20. Use Cramer's Rule to solve the linear system.

$$\begin{cases} x - y + z = -3 \\ 3x + 2y - z = 15 \\ 2x + 3y + z = 0 \end{cases}$$

20. ____________________

Chapter 4 Test

Form C

(Page 1 of 4 pages)

Name ______________________

Date ______________________

1. State the order of the matrix. $\begin{bmatrix} 4 & 1 & 8 & 2 & 9 \end{bmatrix}$ 1. ______________

In 2–6, perform the matrix operations, if possible.

2. $\begin{bmatrix} 2 & 0 & 7 \\ 1 & -3 & 4 \\ -3 & 2 & 4 \end{bmatrix} + \begin{bmatrix} 1 & 9 & -1 \\ 4 & -3 & 2 \\ 6 & -5 & 1 \end{bmatrix}$ 2. ______________

3. $\begin{bmatrix} 4 & 7 & 1 \\ 2 & 0 & -4 \end{bmatrix} - \begin{bmatrix} -1 & -2 & 0 \\ 1 & -4 & 3 \end{bmatrix}$ 3. ______________

4. $-8\begin{bmatrix} 1 & 0 & 6 \\ 2 & 4 & -3 \end{bmatrix}$ 4. ______________

5. $\begin{bmatrix} 1 & 0 \\ 2 & 4 \end{bmatrix}\begin{bmatrix} 1 & 3 \\ 2 & 2 \\ 4 & 1 \end{bmatrix}$ 5. ______________

6. $\begin{bmatrix} 1 & 2 & -1 \\ 3 & -2 & 4 \end{bmatrix}\begin{bmatrix} 1 & -2 \\ 2 & -1 \\ 7 & 3 \end{bmatrix}$ 6. ______________

7. Solve for x and y. $2\begin{bmatrix} x & -3 \\ 1 & 2 \end{bmatrix} - 3\begin{bmatrix} 4 & -2 \\ -1 & y \end{bmatrix} = \begin{bmatrix} -4 & 0 \\ 5 & -5 \end{bmatrix}$ 7. ______________

8. Evaluate $\begin{vmatrix} 10 & 7 \\ -5 & 12 \end{vmatrix}$.   8. ______________

Name ____________________

9. Evaluate the determinant of the matrix. $\begin{bmatrix} 2 & -1 & 4 \\ 3 & 2 & 0 \\ -1 & 4 & 3 \end{bmatrix}$ **9.** ____________

10. Find the inverse of $\begin{bmatrix} 4 & 7 \\ 9 & 8 \end{bmatrix}$. **10.** ____________

11. ***Geometry*** Find the area of the triangle. **11.** ____________

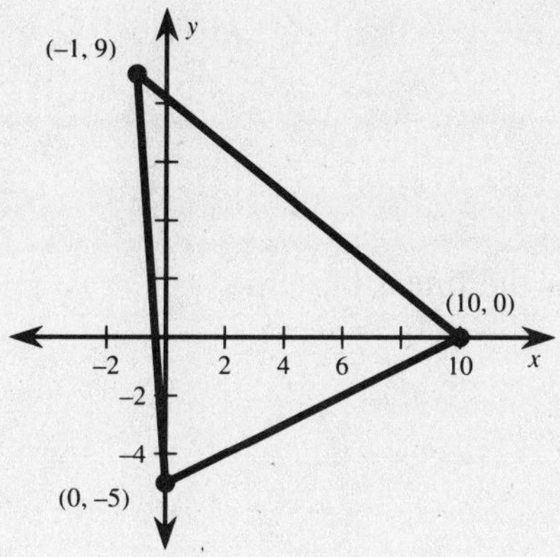

12. ***Inventory*** A company stocks items A, B, and C at each of its two stores. Use matrix multiplication to determine the value of the inventory at each store. **12.** ____________

Inventory

	Item *A*	Item *B*	Item *C*
Store *1*	10	12	15
Store *2*	9	11	8

Value of Each Item ($)

Item *A*	12
Item *B*	15
Item *C*	20

Inventory Value ($)

Store *1*	
Store *2*	

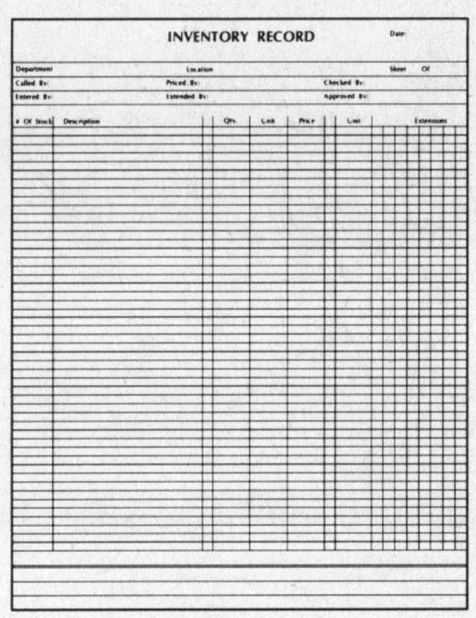

13. Write the linear system as a matrix equation. **13.** ________________

$$\begin{cases} 12x - 37y = 14 \\ 9x + 19y = -7 \end{cases}$$

14. Use an inverse matrix to solve the linear system. **14.** ________________

$$\begin{cases} 2x - 5y = 7 \\ x + 4y = -5 \end{cases}$$

15. Write the augmented matrix for the linear system. **15.** ________________

$$\begin{cases} x - 4y + 2z = 3 \\ 5x - 7y + z = 4 \\ 3x + 2y - 2z = 8 \end{cases}$$

16. Write the linear system which is represented by the augmented matrix. **16.** ________________

$$\left[\begin{array}{ccc|c} 1 & -3 & 7 & 10 \\ 0 & 1 & -4 & 8 \\ 0 & 0 & 1 & -12 \end{array}\right]$$

17. Use an augmented matrix to solve the system. **17.** ________________

$$\begin{cases} x + y + z = -1 \\ 2x - y + 3z = 5 \\ 3x - 2y + 4z = 6 \end{cases}$$

18. Find the coefficient matrix and evaluate its determinant. **18.** ____________

$$\begin{cases} 9x - 2y = 15 \\ 3x + 6y = 5 \end{cases}$$

19. Use Cramer's Rule to solve the linear system. **19.** ____________

$$\begin{cases} 3x - 5y = -2 \\ 4x + 3y = 5 \end{cases}$$

20. Use Cramer's Rule to solve the linear system. **20.** ____________

$$\begin{cases} 3x + 2y = 0 \\ 2x - y + 2z = 3 \\ \phantom{3x + {}} 5y + 6z = 0 \end{cases}$$

Mid-Chapter 5 Test Form A

Name ____________________

Date ____________________

1. Solve the equation $3x^2 - 7 = 8$. **1.** ____________

2. ***Geometry*** Use the Pythagorean Theorem to find x to two decimal places. **2.** ____________

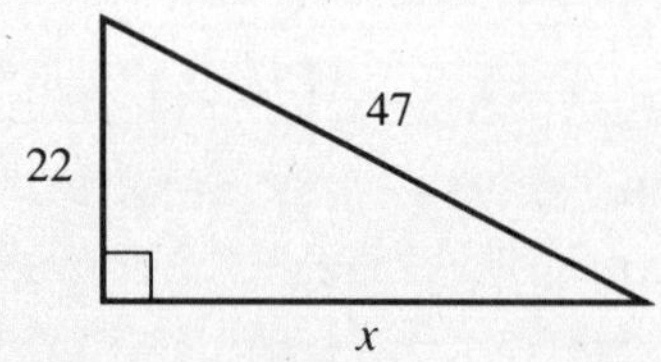

3. For the parabola find the vertex and the axis of symmetry, and determine whether it opens up or down. $y = -4x^2 + 8x + 2$ **3.** ____________

4. Sketch the graph of the equation.

$y = x^2 - x - 2$

4. *Use graph at left.*

5. Solve the equation by completing the square. $x^2 - 4x - 3 = 0$ **5.** ____________

6. ***Geometry*** The base of a triangle is 2 inches longer than the height. Its area is 12 square inches. Find the base and height. **6.** ____________

7. ***Skydiver*** A skydiver jumps from a plane at an altitude of 6000 feet. What is the approximate altitude of the diver after 10 seconds of free fall? **7.** ____________

Mid-Chapter 5 Test Form B

Name ______________________

Date ______________________

1. Solve the equation $5x^2 + 4 = 19$. 1. ______________

2. ***Geometry*** Use the Pythagorean Theorem to find x to two decimal places. 2. ______________

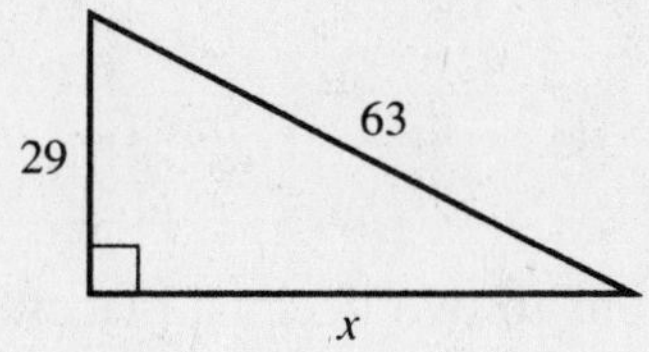

3. For the parabola find the vertex and the axis of symmetry, and determine whether it opens up or down. $y = -2x^2 + 8x - 4$ 3. ______________

4. Sketch the graph of the equation. 4. *Use graph at left.*

   $y = \frac{1}{4}x^2 - 1$

5. Solve the equation by completing the square. $x^2 - 6x - 1 = 0$ 5. ______________

6. ***Geometry*** The area of a rectangle is 45 square inches and the length is 4 inches more than the width. Find the length and width. 6. ______________

7. ***Skydiver*** A skydiver jumps from a plane at an altitude of 8000 feet. What is the approximate altitude of the diver after 12 seconds of free fall? 7. ______________

Name ____________________

Date ____________________

1. Solve the equation.

$-x^2 + 4 = 2x^2 - 5$

1. ____________________

2. Solve the equation. Round the solutions to two decimal places.

$5x^2 - 2 = 7$

2. ____________________

3. ***Geometry*** The surface area of a cube is 536 square inches. How long is each edge? (Round to two decimal places.)

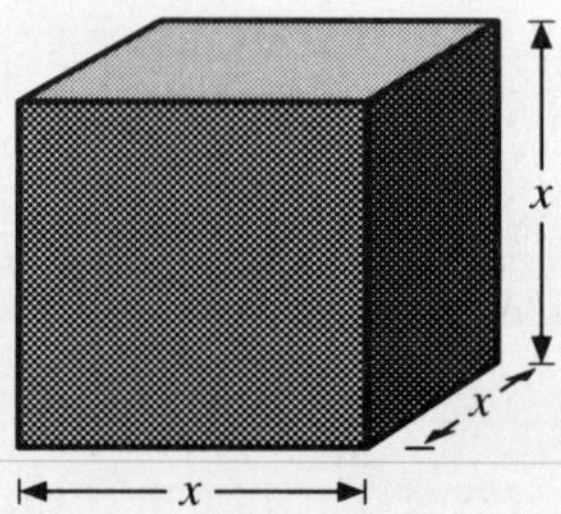

3. ____________________

4. ***Falling Object*** The height, h (in feet), of a falling object on Mars is given by $h = -6t^2 + s$, where t is the time in seconds and s is the initial height in feet. If an object were dropped from a height of 200 feet, how long would it take to reach the ground? (Round to two decimal places.)

4. ____________________

5. Does the parabola open *up* or *down*?

$y = 4 + 6x - 2x^2$

5. ____________________

Name ______________________

6. Find the *vertex* and the *axis of symmetry* of the parabola. **6.** ______________

$y = 3x^2 + 12x + 9$

7. Sketch the graph of the equation.

$y = x^2 - 2x + 3$

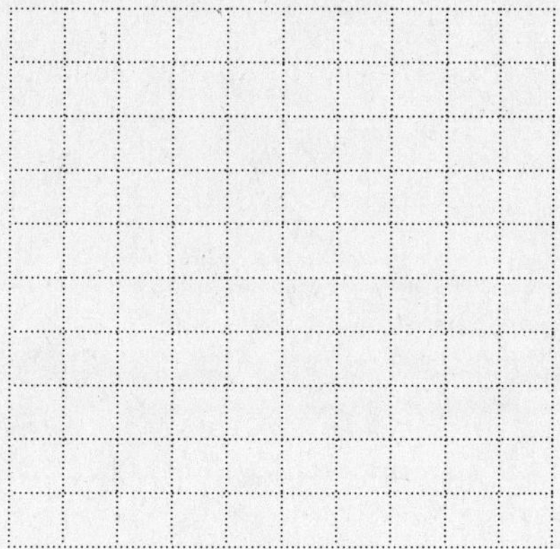

7. *Use graph at left.*

8. Write the trinomial as the square of a binomial. **8.** ______________

$x^2 - 18x + 81$

9. Solve the equation by completing the square. **9.** ______________

$x^2 + 2x - 35 = 0$

10. ***Geometry*** The height of a triangle is three feet longer than the base. The area of the triangle is 35 square feet. Find the height and base of the triangle. **10.** ______________

11. State the discriminant of the quadratic. **11.** ____________

$5x^2 - 3x - 12 = 0$

12. Use the discriminant to determine the number of real solutions of the equation. **12.** ____________

$4x^2 - 3x - 7 = 0$

13. Use the quadratic formula to solve the equation. **13.** ____________

$x^2 + 2x - 1 = 0$

14. Solve the equation. Round to two decimal places. **14.** ____________

$0.2x^2 + 0.31x - 0.15 = 0$

15. Write the number using the imaginary unit i. **15.** ____________

$\sqrt{-36}$

16. Simplify the expression. **16.** ____________

$(3i)^2$

17. Is $-2i$ a solution of $x^2 = -4$? **17.** ____________

Name ______________________

18. Solve the equation.

$4 - 2x^2 = 12$

18. ______________________

19. Perform the indicated operations.

$(5 - 2i) - 2(3 + i)$

19. ______________________

20. Perform the indicated operations.

$(2 + 3i)(1 - 4i)$

20. ______________________

21. Solve the equation.

$2x^2 + x + 3 = 0$

21. ______________________

22. Sketch the graph of the inequality.

$y \geq 2x^2 + 4x - 1$

22. *Use graph at left.*

Chapter 5 Test

Form B

(Page 1 of 4 pages)

Name ______________________

Date ______________________

1. Solve the equation.

$x^2 - 7 = 14 - 2x^2$

1. ____________

2. Solve the equation. Round the solutions to two decimal places.

$7x^2 - 2 = 23$

2. ____________

3. ***Geometry*** The surface area of a cube is 380 square inches. How long is each edge? (Round to two decimal places.)

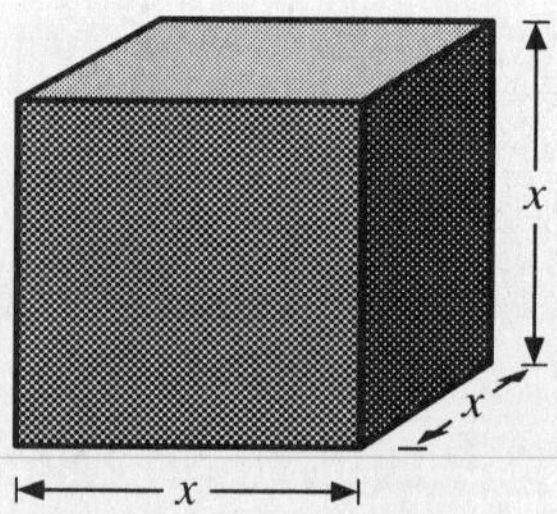

3. ____________

4. ***Falling Object*** The height, h (in feet), of a falling object on Mars is given by $h = -6t^2 + s$, where t is the time in seconds and s is the initial height in feet. If an object were dropped from a height of 125 feet, how long would it take to reach the ground? (Round to two decimal places.)

4. ____________

5. Does the parabola open *up* or *down*?

$y = -7 - 5x + 3x^2$

5. ____________

Name ______________________

6. Find the *vertex* and the *axis of symmetry* of the parabola.

$y = -3x^2 + 12x - 8$

6. ______________________

7. Sketch the graph of the equation.

$y = x^2 - 4x + 4$

7. *Use graph at left.*

8. Write the trinomial as the square of a binomial.

$x^2 - \frac{2}{3}x + \frac{1}{9}$

8. ______________________

9. Solve the equation by completing the square.

$x^2 + 2x - 24 = 0$

9. ______________________

10. ***Geometry*** The height of a triangle is three feet longer than the base. The area of the triangle is 44 square feet. Find the height and base of the triangle.

10. ______________________

 Name ______________________

11. State the discriminant of the quadratic. 11. ____________

$3x^2 - x + 4 = 0$

12. Use the discriminant to determine the number of real solutions of the equation. 12. ____________

$5x^2 - 3x + 1 = 0$

13. Use the quadratic formula to solve the equation. 13. ____________

$3x^2 + x - 1 = 0$

14. Solve the equation. Round to two decimal places. 14. ____________

$0.37x^2 - 0.18x - 1.23 = 0$

15. Write the number using the imaginary unit i. 15. ____________

$\sqrt{-49}$

16. Simplify the expression. 16. ____________

$3i^2$

17. Is $-3i$ a solution of $x^2 = -3$? 17. ____________

 Name ____________________

18. Solve the equation.

$$4x^2 + 20 = 0$$

18. ____________________

19. Perform the indicated operations.

$$(5 - 2i) + (3 - 2i)$$

19. ____________________

20. Perform the indicated operations.

$$(-3 + 7i)(1 - 2i)$$

20. ____________________

21. Solve the equation.

$$2x^2 - x + 2 = 0$$

21. ____________________

22. Sketch the graph of the inequality.

$$y \leq -2x^2 + 8x - 6$$

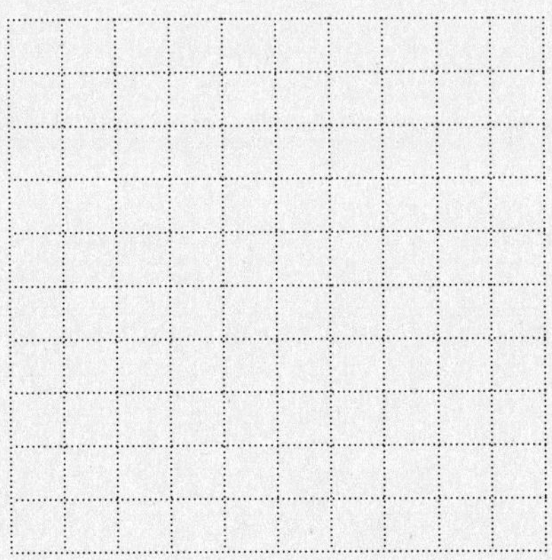

22. *Use graph at left.*

Chapter 5 Test

Form C Name ____________

(Page 1 of 4 pages) Date ____________

1. Solve the equation. **1.** ____________

   $\frac{1}{3}x^2 + 1 = 33$

2. Solve the equation and write the answer to two decimal places. **2.** ____________

   $-\frac{1}{2}x^2 + 5 = \frac{1}{6}$

3. ***Geometry*** The surface area of a sphere is given by $S = 4\pi r^2$ where r is the radius and $\pi \approx 3.14$. Find the radius of a sphere whose surface area is 748 square centimeters. **3.** ____________

4. ***Falling Object*** How long, in seconds, does it take an object dropped from 200 feet to strike the ground? Write the result to two decimal places. **4.** ____________

5. Find the *vertex* of the parabola and determine if it opens *up* or *down*. **5.** ____________

   $y = 7 - 8x - 2x^2$

Name ______________________

6. Find the axis of symmetry and the x-intercepts of the parabola.

$y = -3x^2 + 6x$

6. ______________________

7. Sketch the graph of the equation.

$y = -x^2 - 4x - 1$

7. *Use graph at left.*

8. Write the trinomial as the square of a binomial.

$x^2 - \frac{6}{5}x + \frac{9}{25}$

8. ______________________

9. Solve the equation by completing the square.

$x^2 - 4x - 2 = 0$

9. ______________________

10. ***Geometry*** The base of a triangle is five feet longer than the height. The area of the triangle is 75 square feet. Find the height and base of the triangle.

10. ______________________

Name ______________________

11. State the discriminant of the quadratic. **11.** ____________

$10x^2 - 7x - 3 = 0$

12. Use the discriminant to determine the number of real solutions of the equation. **12.** ____________

$9x^2 - 30x + 25 = 0$

13. Use the quadratic formula to solve the equation. **13.** ____________

$x^2 + x - 8 = 0$

14. Solve the equation. Round to two decimal places. **14.** ____________

$0.1x^2 + 0.01x - 0.001 = 0$

15. Write the number using the imaginary unit i. **15.** ____________

$-\sqrt{-81}$

16. Simplify the expression. **16.** ____________

$\sqrt{3}\sqrt{-3}$

17. Is $-2i\sqrt{2}$ a solution of $x^2 = -8$? **17.** ____________

Name ______________________

18. Solve the equation.

$$4x^2 + 5 = -7$$

18. ______________________

19. Perform the indicated operations.

$$-i + (7 - 5i) - 3(2 - 3i)$$

19. ______________________

20. Perform the indicated operations.

$$(2i)(1 - 4i)(1 + i)$$

20. ______________________

21. Solve the equation.

$$x^2 - x + 2 = 0$$

21. ______________________

22. Sketch the graph of the inequality.

$$y > \tfrac{1}{2}x^2 - 2x - 2$$

22. *Use graph at left.*

Mid-Chapter 6 Test Form A

Name ______________________

Date ______________________

1. Determine whether the relation is a function. **1.** ______

   $(-3,\ 1), (-2,\ 1), (-1,\ 0), (0,\ 0), (1,\ -1), (2,\ -1)$

2. Find $f\left(\frac{3}{2}\right)$. **2.** ______

   $f(x) = -4x^2 + 6x - 2$

3. Find the domain and range of the function. **3.** ______

   $f(x) = \sqrt{x+1}$

4. Find $f(x) \cdot g(x)$ if $f(x) = 2x - 1$ and $g(x) = 2 - x$. **4.** ______

5. Find $f(g(x))$ if $f(x) = 2x - 1$ and $g(x) = 2 - x$. **5.** ______

6. ***Sales Tax*** The sales tax on an item is 6%. Let x represent the price of the item and let $f(x)$ represent the total amount (including sales tax). Write an equation for $f(x)$. **6.** ______

7. Write an equation for the inverse of the relation. **7.** ______

   $y = -7x + 6$

8. Sketch the graph of the function. Is the inverse of $f(x)$ a function? **8.** ______ *Use graph at left.*

   $f(x) = 2 - |x|$

Mid-Chapter 6 Test Form B

Name ______________________

Date ______________________

1. Determine whether the relation is a function. 1. ____________

   $(-1,\ 2), (-1,\ 1), (0,\ 0), (0,\ -1), (1,\ -2), (1,\ -3)$

2. Find $f(-\frac{3}{2})$. 2. ____________

   $f(x) = -4x^2 + 6x - 2$

3. Find the domain and range of the function. 3. ____________

   $f(x) = -\sqrt{3 - x}$

4. Find $2g(x) - f(x)$ if 4. ____________
   $f(x) = 1 - 3x$ and $g(x) = x - 3$.

5. Find $g(f(x))$ if 5. ____________
   $f(x) = 1 - 3x$ and $g(x) = x - 3$.

6. ***Sales Tax*** The sales tax on an item is 7%. Let x represent the price of an item and let $f(x)$ represent the total amount (including sales tax). Write an equation for $f(x)$. 6. ____________

7. Write an equation for the inverse of the relation. 7. ____________

   $y = 13x - 8$

8. Sketch the graph of the function. Is the inverse of $f(x)$ a function? 8. ____________ *Use graph at left.*

   $f(x) = |x - 1|$

Chapter 6 Test

Form A

(Page 1 of 4 pages)

Name ______________________

Date ______________________

1. Determine whether the relation is a function.

$(0,\ 4), (1,\ 4), (2,\ 5), (3,\ 6), (4,\ 6)$

1. ______________

2. Find $f(\frac{2}{3})$.

$f(x) = 18x^2 - 12x - 3$

2. ______________

3. ***Geometry*** The surface area of a sphere with radius r is given by $f(r) = 4\pi r^2$. Find $f(\frac{3}{2})$.

3. ______________

4. Find the domain and range of the function.

$$f(x) = \frac{2}{\sqrt{x-4}}$$

4. ______________

In 5 and 6, $f(x) = 3 - 2x$ and $g(x) = 4x + 1$.

5. $h(x) = g(x) \div f(x)$. Write the equation for $h(x)$.

5. ______________

6. Find $g(f(x))$.

6. ______________

7. What is the domain of $h(x)$ in Problem 5?

7. ______________

8. Find the inverse of the relation.

$(1,\ 7), (2,\ 5), (3,\ 3), (4,\ 1)$

8. ______________

Name ____________________

9. Write an equation for the inverse of the relation.

$y = -11x + 9$

9. ____________________

10. Are f and g inverses of each other?

$g(x) = \frac{1}{2}x - \frac{1}{3}, \; f(x) = \dfrac{6x + 2}{3}$

10. ____________________

11. Sketch the graph of the function. Is the inverse of $f(x)$ a function?

$f(x) = 4 - x^2$

11. ____________________
Use graph at left.

12. Sketch the graph of the function and its inverse on the same coordinate plane.

$f(x) = 2 - 2x$

12. *Use graph at left.*

13. Evaluate $f(-3)$. $f(x) = \begin{cases} -x^2 + 2x, & x \leq 1 \\ -2x + 3, & x > 1 \end{cases}$

13. ____________________

14. Write $f(x) = |x - 4|$ as a compound function.

14. ____________________

15. Sketch the graph of the function.

$$f(x) = \begin{cases} -x, & x \leq 0 \\ x^2, & x > 0 \end{cases}$$

15. *Use graph at left.*

16. Sketch the graph of the step function.

$$f(x) = \begin{cases} 5, & 0 \leq x < 5 \\ 10, & 5 \leq x < 10 \\ 15, & 10 \leq x < 15 \end{cases}$$

16. *Use graph at left.*

17. How is the graph of $f(x) = x^2 - 5$ obtained from the graph of $g(x) = x^2$?

17. ____________________

18. Sketch the graph of the function.

$$f(x) = |x + 3|$$

18. *Use graph at left.*

Name ______________________

19. Find the first five values of the recursive function.

$f(1) = 5;\ f(n) = f(n-1) - n$

19. ______________________

20. ***Shirt Prices*** The data below represents the price of a shirt at ten stores. Find the *mean*, *median*, and *mode* of the data.

48, 33, 29, 52, 37, 44, 29, 35, 44, 29

20. ______________________

Chapter 6 Test

Form B

(Page 1 of 4 pages)

Name ____________________

Date ____________________

1. Determine whether the relation is a function. 1. ____________

   (4, 0), (4, 1), (5, 2), (6, 3), (6, 4)

2. Find $f(-\frac{2}{3})$. 2. ____________

   $f(x) = 18x^2 - 12x - 3$

3. ***Geometry*** The surface area of a sphere with radius r is given by $f(r) = 4\pi r^2$. Find $f(\frac{2}{3})$. 3. ____________

4. Find the domain and range of the function. 4. ____________

   $$f(x) = \frac{-2}{\sqrt{x-3}}$$

In 5 and 6, $f(x) = 3 - 2x$ and $g(x) = 4x + 1$.

5. $h(x) = f(x) \div g(x)$. Write the equation for $h(x)$. 5. ____________

6. Find $f(g(x))$. 6. ____________

7. What is the domain of $h(x)$ in Problem 5? 7. ____________

8. Find the inverse of the relation. 8. ____________

   (1, 5), (2, 3), (3, 1), (4, −1)

Name ______________________

9. Write an equation for the inverse of the relation.

$y = 18x - 7$

9. ______________________

10. Are f and g inverses of each other?

$g(x) = \frac{1}{3}x - \frac{1}{2}, \ f(x) = \frac{3}{2}(2x + 1)$

10. ______________________

11. Sketch the graph of the function. Is the inverse of $f(x)$ a function?

$f(x) = 2 - x^2$

11. ______________________
Use graph at left.

12. Sketch the graph of the function and its inverse on the same coordinate plane.

$f(x) = 2 - 4x$

12. *Use graph at left.*

13. Evaluate $f(-3)$. $f(x) = \begin{cases} x, & x \le 0 \\ x^2 - 3x, & x > 0 \end{cases}$

13. ______________________

14. Write $f(x) = |2x + 3|$ as a compound function.

14. ______________________

Name ____________________

15. Sketch the graph of the function.

$$f(x) = \begin{cases} -\frac{1}{2}x - 1, & x \leq 0 \\ 2x - 1, & x > 0 \end{cases}$$

15. *Use graph at left.*

16. Sketch the graph of the step function.

$$f(x) = \begin{cases} 0, & 0 < x \leq 5 \\ 5, & 5 < x \leq 10 \\ 15, & 10 < x \leq 15 \end{cases}$$

16. *Use graph at left.*

17. How is the graph of $f(x) = x^2 + 2$ obtained from the graph of $g(x) = x^2 - 2$?

17. ____________________

18. Sketch the graph of the function.

$$f(x) = |2 - x|$$

18. *Use graph at left.*

Name ______________________

19. Find the first five values of the recursive function.

$f(1) = 5;\ f(n) = f(n-1) + n$

19. ______________________

20. ***Shoe Prices*** The data below represents the price of a pair of shoes at ten stores. Find the *mean*, *median*, and *mode* of the data.

77, 65, 59, 77, 63, 60, 60, 61, 68, 77

20. ______________________

Chapter 6 Test

Form C

(Page 1 of 4 pages)

Name ______________________

Date ______________________

1. Determine whether the relation is a function. **1.** __________

$(-3, 3), (-2, 2), (-1, 1), (1, -1), (2, -2), (3, -3)$

2. Find $f(-\frac{1}{3})$. **2.** __________

$f(x) = 18x^2 - 12x - 3$

3. ***Geometry*** The volume of a sphere with radius r is given by $f(r) = \frac{4}{3}\pi r^3$. Find $f(\frac{3}{2})$. **3.** __________

4. Find the domain and range of the function. **4.** __________

$$f(x) = \frac{\sqrt{x+1}}{x^2}$$

In 5 and 6, $f(x) = x^2 + 4$ and $g(x) = 2x - 3$.

5. $h(x) = f(x) \div g(x)$. Write the equation for $h(x)$. **5.** __________

6. Find $g(f(x))$. **6.** __________

7. What is the domain of $h(x)$ in Problem 5? **7.** __________

8. Find the inverse of the relation. **8.** __________

$(1, 1), (2, 2), (1, 3), (4, 4)$

Name ____________________

9. Write an equation for the inverse of the relation.

$y = -22x - 7$

9. ____________________

10. Are f and g inverses of each other?

$f(x) = \frac{2}{3}x - \frac{1}{2},\ g(x) = \frac{3}{2}x - \frac{3}{4}$

10. ____________________

11. Sketch the graph of the function. Is the inverse of $f(x)$ a function?

$f(x) = -x^2 - 1$

11. ____________________ *Use graph at left.*

12. Sketch the graph of the function and its inverse on the same coordinate plane.

$f(x) = \frac{2}{3}x - 2$

12. *Use graph at left.*

13. Evaluate $f(-3)$. $f(x) = \begin{cases} -x^2 - 4x, & x \leq 0 \\ x, & x > 0 \end{cases}$

13. ____________________

14. Write $f(x) = |3 - 5x|$ as a compound function.

14. ____________________

15. Sketch the graph of the function.

$$f(x) = \begin{cases} -\frac{1}{2}x, & x \le 2 \\ (x-2)^2 - 1, & x > 2 \end{cases}$$

15. *Use graph at left.*

16. Sketch the graph of the step function.

$$f(x) = \begin{cases} 29, & 0 < x \le 1 \\ 52, & 1 < x \le 2 \\ 75, & 2 < x \le 3 \end{cases}$$

16. *Use graph at left.*

17. How is the graph of $f(x) = (x+2)^2$ obtained from the graph of $g(x) = x^2$?

17. ______________________

18. Sketch the graph of the function.

$$f(x) = 3 - |x - 2|$$

18. *Use graph at left.*

Name ____________________

19. Find the first five values of the recursive function. **19.** ____________

$f(1) = 5;\ f(n) = n - f(n-1)$

20. ***Dress Prices*** The data below represents the price of a dress at 10 stores. Find the *mean, median*, and *mode* of the data. **20.** ____________

85, 99, 90, 98, 90, 86, 93, 95, 87, 96

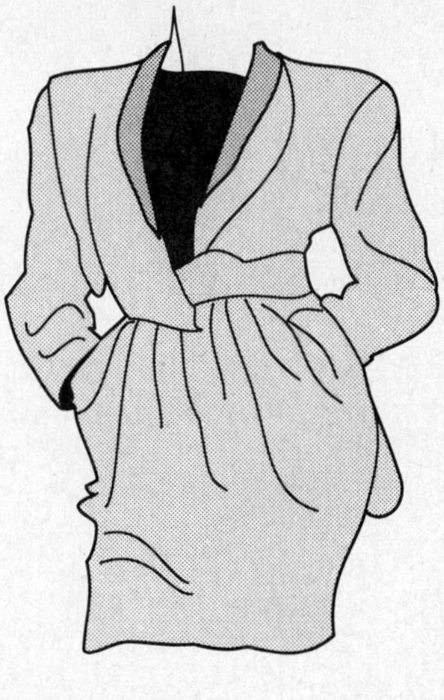

Cumulative Test

Chapters 1–6

(Page 1 of 8 pages)

Name ______________________

Date ______________________

1. Write the numbers in increasing order.

$0, -\frac{3}{5}, \frac{5}{3}, -\frac{2}{3}, \frac{4}{7}, \frac{7}{4}, -\frac{3}{2}.$

1. ______________________

2. State the property that is illustrated.

$(5 \cdot 2) \cdot 3 = 3 \cdot (5 \cdot 2)$

2. ______________________

3. Evaluate $100 - (6)(4x) \div y$ when $x = 20$ and $y = 5$.

3. ______________________

4. Use a calculator to evaluate the expression to two decimal places.

$9.83z - (7.16)(3.82)$ when $z = 6.62$

4. ______________________

5. Solve the equation.

$3(5 + 4x) = 7 - 2(3 - 4x)$

5. ______________________

6. Solve for t.

$A = P(1 + rt)$

6. ______________________

 Name ____________

7. Solve and graph the inequality.

$2x - 5 < 3(x - 2)$

7. ____________ *Use graph at left.*

8. Solve and graph the inequality.

$|3 - 2x| \leq 5$

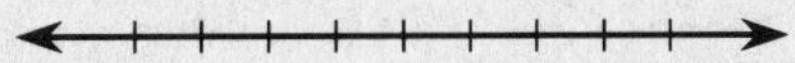

8. ____________ *Use graph at left.*

9. Write the equation of the horizontal line passing through the point $(-7, -2)$.

9. ____________

10. Write the equation of the line passing through the points $(7, 1)$ and $(0, -3)$ in slope-intercept form.

10. ____________

11. Which of the given lines are parallel?

$2x - 4y = 5, \; 4x + 2y = 5, \; x - 2y = 5$

11. ____________

Name ____________

12. ***Bicycling*** At 12 noon two bicyclists going in the same direction are 15 miles apart. At 2:30 P.M. they are 10 miles apart. Write an equation for the distance, D (in miles), separating the bicyclists in terms of the time t. (Assume $t = 0$ at 12 noon.)

12. ____________

13. The values of x and y vary *directly*. $y = 18$ when $x = 4$. Write an equation that relates the variables.

13. ____________

14. Is the ordered pair (6, 7) a solution of the inequality?

$$5x - 3y \leq 7$$

14. ____________

15. Sketch the graph of the inequality.

$$y \geq \tfrac{3}{2}x + 3$$

15. *Use graph at left.*

 Name ____________________

16. Evaluate $5 - 3|2 - 4x|$ when $x = 2$.

16. ____________________

17. Sketch the graph of the equation.

$y = 3 - |2x - 1|$

Label the vertex.

17. *Use graph at left.*

18. Are there none, one, or many solutions to the system?

$$\begin{cases} \frac{1}{2}x - 3y = 2 \\ 2x - 12y = 8 \end{cases}$$

18. ____________________

19. Solve the system of linear equations.

$$\begin{cases} 3x + 6y = 4 \\ 2x - 3y = -4 \end{cases}$$

19. ____________________

20. Graph the system of linear inequalities and find the maximum value of C under the constraints.

$C = -4x + 7y$ Constraints

$$\begin{cases} y \leq \frac{1}{2}x + 1 \\ y \geq -1 \\ x \leq 3 \end{cases}$$

20. ____________________
Use graph at left.

 Name ____________

21. Perform the matrix operations, if possible. **21.** ____________

$$2\begin{bmatrix} 1 & 7 \\ -1 & 4 \end{bmatrix} - 3\begin{bmatrix} 2 & 0 \\ 5 & -2 \end{bmatrix}$$

22. Perform the matrix operations, if possible. **22.** ____________

$$\begin{bmatrix} 3 & -1 & 4 \\ 5 & 0 & -2 \end{bmatrix}\begin{bmatrix} 1 & 3 \\ 4 & -2 \\ -2 & 0 \end{bmatrix}$$

23. Solve for x and y. **23.** ____________

$$\begin{bmatrix} 4 & -1 \\ 5 & 3 \end{bmatrix} - 2\begin{bmatrix} x & 1 \\ -1 & y \end{bmatrix} = \begin{bmatrix} 8 & -3 \\ 7 & 9 \end{bmatrix}$$

24. Find the inverse of $\begin{bmatrix} 3 & -7 \\ 8 & 4 \end{bmatrix}$. **24.** ____________

25. Evaluate $\begin{vmatrix} 4 & 8 \\ -7 & -5 \end{vmatrix}$. **25.** ____________

26. Evaluate the determinant of the matrix. **26.** ____________

$$\begin{bmatrix} 3 & -2 & 1 \\ -2 & 4 & -3 \\ 4 & -1 & 1 \end{bmatrix}$$

 Name ______________________

27. Use an augmented matrix to solve the system.

$$\begin{cases} x + y + z = -2 \\ 3x - y + 2z = 1 \\ 2x - 2y - z = 9 \end{cases}$$

27. ______________________

28. Use Cramer's Rule to solve the linear system.

$$\begin{cases} 5x - 4y = 8 \\ 2x + 7y = -3 \end{cases}$$

28. ______________________

29. Solve the equation.

$5x^2 - 8 = 22$

29. ______________________

30. Solve the equation.

$3x^2 + 5x - 1 = 0$

30. ______________________

31. Sketch the graph of the parabola. Label the vertex.

$y = -x^2 + 4x - 1$

31. *Use graph at left.*

32. Solve the equation. Round your result to two decimal places. **32.** ____________

$0.3x^2 - 0.21x - 0.14 = 0$

33. Simplify $(2i)^2 - (3i)(i^2)$. **33.** ____________

34. Perform the indicated operations and simplify. **34.** ____________

$(3 - 2i)(4 - i) + (-2 - 5i)$

35. Solve the equation. **35.** ____________

$4x^2 + x + 1 = 0$

36. Is the relation a function? If so, find its inverse. **36.** ____________

$(-2,\ 4), (-1,\ 3), (1,\ 2), (2,\ 1)$

37. $f(x) = 1 - 3x;\ g(x) = \frac{1}{3}x + 1.$ **37.** ____________

Find $f(x) \cdot g(x)$ and $f(g(x))$.

 Name ______________________

38. Find the inverse of the function and graph both on the same coordinate axes.

$f(x) = -\frac{1}{3}x - 1$

38. *Use graph at left.*

39. Evaluate $f(3)$.

$$f(x) = \begin{cases} x^2 - 3, & x \le 2 \\ 3x - 5, & x > 2 \end{cases}$$

39. ______________________

40. Write $f(x) = |3x - 6|$ as a compound function.

40. ______________________

41. How is the graph of $f(x) = (x - 1)^2$ obtained from the graph of $g(x) = x^2$?

41. ______________________

42. Find the first four values of the recursive function.

$f(1) = 3;\ f(n) = f(n - 1) + 2n$

42. ______________________

43. Find the mean, median, and mode of the data given below.

10, 5, 8, 7, 7, 9, 6, 11, 7, 10, 12

43. ______________________

Mid-Chapter 7 Test Form A

Name ____________________

Date ____________________

1. Simplify the expression. $\dfrac{(x^2y^2)^{-1}}{xy^{-3}}$ 1. ________

2. Solve for x. $\dfrac{a^3}{a^{-x}} = a^2$ 2. ________

3. ***Balance in an Account*** $4750 is deposited in an account paying 5.5% annual interest, compounded quarterly. Find the balance after 5 years. 3. ________

4. Does $y = 2(0.95)^x$ serve as a model for exponential growth or exponential decay? 4. ________

5. Rewrite $\sqrt[7]{20}$ using rational exponents. 5. ________

6. Use radical notation to write the fourth roots of 23. 6. ________

7. Evaluate $\sqrt[6]{1048}$ to three decimal places. 7. ________

8. ***Population Model*** Starting in 1980, a district lost 10% of its population each year for the next 5 years. What percent of its 1980 population was left in 1985? 8. ________

Mid-Chapter 7 Test Form B

Name ____________________

Date ____________________

1. Simplify the expression. $\dfrac{x^{-1}y^3}{(x^2y^{-1})^{-2}}$ 1. ____________

2. Solve for x. $\dfrac{a^5}{a^{-2x}} = a^8$ 2. ____________

3. ***Balance in an Account*** $3250 is deposited in an account paying 5.5% annual interest, compounded quarterly. Find the balance after six years. 3. ____________

4. Does $y = \frac{1}{3}\left(\frac{5}{4}\right)^x$ serve as a model for exponential growth or exponential decay? 4. ____________

5. Rewrite $\sqrt[6]{17}$ using rational exponents. 5. ____________

6. Use radical notation to write the fourth roots of 77. 6. ____________

7. Evaluate $\sqrt[6]{1491}$ to three decimal places. 7. ____________

8. ***Population Model*** Starting in 1980, a district lost 10% of its population each year for the next 6 years. What percent of its 1980 population was left in 1986? 8. ____________

Chapter 7 Test

Form A

(Page 1 of 3 pages)

Name ______________________

Date ______________________

1. Simplify. $(-3x^{-2})^3$ — 1. ____________

2. Simplify. $\frac{6x^2}{y^3} \cdot \frac{y^{-2}x^3}{9x^2}$ — 2. ____________

3. Solve for x. $3^3 \cdot 3^x \cdot 3^{x-1} = 3^{12}$ — 3. ____________

4. ***Balance in an Account*** Find the value of \$1000 deposited for 10 years in an account paying 7% annual interest compounded yearly. — 4. ____________

5. ***Deposit in an Account*** How much money must be deposited now in an account paying 8% annual interest, compounded quarterly, to have a balance of \$1000 after 10 years? — 5. ____________

6. ***Depreciation Value*** A piece of equipment costs \$85,000 new but depreciates 15% per year in each succeeding year. Find its value after 10 years. — 6. ____________

Name ______________________

7. Evaluate. $16^{5/4}$ **7.** ______________

8. Use a calculator to evaluate $9^{-1/3}$ to three decimal places. **8.** ______________

9. ***Geometry*** The volume of a dodecahedron is $V \approx 7.66312a^3$ where a is the length of an edge. Find the edge length of a dodecahedron whose volume is 1000 cubic centimeters. **9.** ______________

10. Rewrite $7^{1/5}$ using radical notation. **10.** ______________

11. Evaluate $\sqrt[5]{1540}$ to three decimal places using a calculator. **11.** ______________

12. Simplify. $\dfrac{25^{1/6}}{25^{2/3}}$ **12.** ______________

13. Simplify. $\sqrt[3]{40} + 4\sqrt[3]{5}$ **13.** ______________

14. Simplify. $(5^{2/9})^{3/4}$ **14.** ______________

15. Solve the equation. $\sqrt[3]{y-2}=5$ **15.** ______________

16. The geometric mean of 10 and x is $5\sqrt{2}$. Find x. **16.** ______________

17. Find the distance between the points. **17.** ______________

$(-1,\ 4)$ and $(3,\ 1)$

In 18–20, refer to the function $g(x) = 2 + \sqrt{x+1}$.

18. What is the domain of $g(x)$? **18.** ______________

19. What is the range of $g(x)$? **19.** ______________

20. Sketch the graph of $g(x)$.

20. *Use graph at left.*

Chapter 7 Test

Form B

(Page 1 of 3 pages)

Name ______________________

Date ______________________

1. Simplify. $(-2x^{-3})^2$

1. ______________________

2. Simplify. $\dfrac{4x^3}{y^2} \cdot \dfrac{y^{-3}x^{-2}}{8x^{-1}}$

2. ______________________

3. Solve for x. $4^{-2} \cdot 4^{x+1} \cdot 4^3 = 4^5$

3. ______________________

4. ***Balance in an Account*** Find the value of \$1000 deposited for 8 years in an account paying 8% annual interest compounded semiannually.

4. ______________________

5. ***Deposit in an Account*** How much money must be deposited now in an account paying 7% annual interest, compounded yearly, to have a balance of \$1000 after 6 years?

5. ______________________

6. ***Depreciated Value*** A piece of equipment costs \$75,000 new but depreciates 12% per year in each succeeding year. Find its value after 8 years.

6. ______________________

Name ____________________

7. Evaluate. $(27)^{2/3}$ 7. ________

8. Evaluate $(16)^{-1/3}$ to three decimal places. 8. ________

9. ***Geometry*** The volume of a dodecahedron is $V \approx 7.66312a^3$ where a is the length of an edge. Find the edge length of a dodecahedron whose volume is 600 cubic centimeters. 9. ________

10. Rewrite $18^{1/6}$ using radical notation. 10. ________

11. Evaluate $\sqrt[7]{4983}$ to three decimal places using a calculator. 11. ________

12. Simplify. $\dfrac{49^{5/6}}{49^{1/3}}$ 12. ________

13. Simplify. $\sqrt[4]{32} + 5\sqrt[4]{2}$ 13. ________

Name ____________________

14. Simplify. $(7^{3/8})^{4/9}$ **14.** ____________________

15. Solve the equation. $\sqrt{2y-7}=11$ **15.** ____________________

16. The geometric mean of 20 and x is $4\sqrt{5}$. Find x. **16.** ____________________

17. Find the distance between the points. **17.** ____________________

$(5, -6)$ and $(-3, 0)$

In 18–20, refer to the function $g(x) = 1 + \sqrt{x+3}$.

18. What is the domain of $g(x)$? **18.** ____________________

19. What is the range of $g(x)$? **19.** ____________________

20. Sketch the graph of $g(x)$. **20.** *Use graph at left.*

Chapter 7 Test

Form C

(Page 1 of 3 pages)

Name ______________________

Date ______________________

1. Simplify. $(-3x^{-2})^{-3}$ — 1. ______

2. Simplify. $\dfrac{12x^{-3}}{y^4} \cdot \dfrac{(y^{-2}x^2)^{-1}}{15x^{-2}}$ — 2. ______

3. Solve for x. $2^3 \cdot 4^x \cdot 8^2 = 16^3$ — 3. ______

4. ***Balance in an Account*** Find the value of \$1000 deposited for 10 years in an account paying 6% annual interest compounded monthly. — 4. ______

5. ***Deposit in an Account*** How much money must be deposited now in an account paying 7.25% annual interest, compounded quarterly, to have a balance of \$1000 after 10 years? — 5. ______

6. ***Depreciated Value*** A piece of equipment costs \$95,000 new but depreciates 14% per year in each succeeding year. Find its value after 10 years. — 6. ______

 Name ______________

7. Evaluate. $16^{-5/4}$ **7.** ______________

8. Use a calculator to evaluate $25^{-4/3}$ to three decimal places. **8.** ______________

9. ***Geometry*** The volume of a dodecahedron is $V \approx 7.66312a^3$ where a is the length of an edge. Find the edge length of a dodecahedron whose volume is 975 cubic inches. **9.** ______________

10. Rewrite $11^{1/4}$ using radical notation. **10.** ______________

11. Evaluate $\sqrt[7]{9804}$ to three decimal places using a calculator. **11.** ______________

12. Simplify. $\dfrac{81^{1/12}}{81^{1/3}}$ **12.** ______________

13. Simplify. $\sqrt[4]{32} + \sqrt[4]{162}$ **13.** ______________

Name ______________________

14. Simplify. $(7^{3/4})^{-2/9}$

14. ______________________

15. Solve the equation. $\sqrt{x^2 + 5} = 3 - x$

15. ______________________

16. The geometric mean of $2\sqrt[3]{4}$ and x is 2. Find x.

16. ______________________

17. Find the distance between the points.

$(2, -\frac{9}{5})$ and $(\frac{7}{5}, -1)$

17. ______________________

In 18–20, refer to the function $h(x) = 2 - \sqrt{x + 1}$.

18. What is the domain of $h(x)$?

18. ______________________

19. What is the range of $h(x)$?

19. ______________________

20. Sketch the graph of $h(x)$.

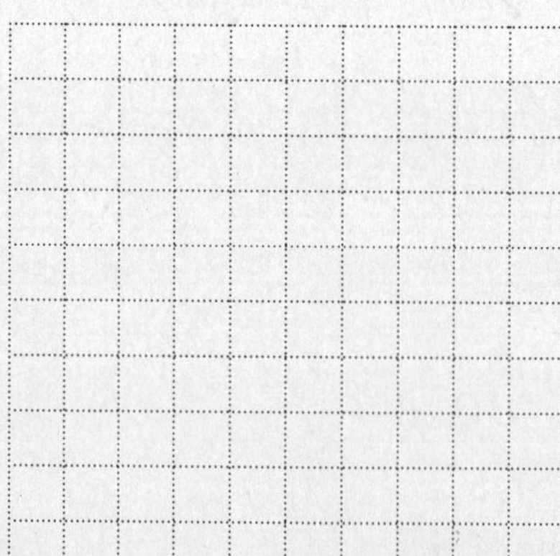

20. *Use graph at left.*

Mid-Chapter 8 Test Form A

Name ______________________

Date ______________________

1. Sketch the graph of
   $f(x) = 2^{x-1}$.

1. *Use graph at left.*

2. Evaluate $6^{2\pi}$ to three decimal places.

2. ______________________

3. ***Depreciated Value*** The value of the new car that you bought in 1990 for \$18,500, decreases by 18% each year. What will the car be worth in 1996?

3. ______________________

4. Write $\log_3 81 = 4$ in exponential form.

4. ______________________

5. Evaluate without using a calculator.
   $\log_4 8$

5. ______________________

6. Evaluate $\log_7 14$ to three decimal places.

6. ______________________

7. Solve for x. $\log_x 64 = 6$

7. ______________________

8. Expand the expression. $\log_6 \dfrac{5x}{y}$

8. ______________________

9. Condense the expression.
   $3 \log_{10} 2 - \log_{10} y + \frac{1}{3} \log_{10} x$

9. ______________________

10. Solve for x.
    $\log_3(x + 1) - \log_3(x) = \log_3 4$

10. ______________________

Mid-Chapter 8 Test Form B

Name ______________________

Date ______________________

1. Sketch the graph of
   $f(x) = -2^x + 3$.

1. *Use graph at left.*

2. Evaluate $5^{2\sqrt{3}}$ to three decimal places.

2. ______________________

3. ***Depreciated Value*** The value of the new car that you bought in 1990 for \$17,800, decreases by 16% each year. What will the car be worth in 1995?

3. ______________________

4. Write the equation in logarithmic form.
   $8^{1.20} \approx 12.126$

4. ______________________

5. Evaluate without using a calculator.
   $\log_3 \frac{1}{9}$

5. ______________________

6. Evaluate $\log_{10} 7$ to three decimal places.

6. ______________________

7. Solve for x. $\log_{16} 64 = x$

7. ______________________

8. Expand the expression. $\log_5 2\sqrt{x}$

8. ______________________

9. Condense the expression.
   $3\log_4 x - \frac{1}{2}\log_4 y + \log_4 7$

9. ______________________

10. Solve for x.
    $\log_4 3 + \log_4(1 - x) = \log_4(1 + x)$

10. ______________________

Chapter 8 Test

Form A

(Page 1 of 3 pages)

Name ______________________

Date ______________________

1. Evaluate $7^{\sqrt{5}/2}$ to three decimal places. **1.** ______________

2. ***Company Sales*** A company had total sales of $2,300,000 in 1980. Each year between 1980 and 1986 the sales increased by 14%. Approximate to the nearest $100,000 the sales for 1986. **2.** ______________

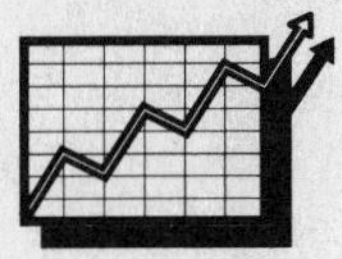

3. Sketch the graph.

$f(x) = \left(\frac{1}{2}\right)^{x+2}$

3. *Use graph at left.*

4. Write $4^{-3} = \frac{1}{64}$ in logarithmic form. **4.** ______________

5. Evaluate without using a calculator. **5.** ______________

$\log_5 5^{0.3}$

6. Which of the following is *undefined*? **6.** ______________

$\log_4 16, \ \log_4 1, \ \log_4(-4)$

7. Evaluate to three decimal places. **7.** ______________

$\log_8 17$

8. Solve for x. $\log_4 \frac{1}{16} = x$ **8.** ________________

9. Expand the expression. $\log_3(x^{-2}y^3)$ **9.** ________________

10. Condense the expression. **10.** ________________

$\frac{1}{2}\log_5 16 - 3\log_5 x + 4\log_5 y$

11. Solve for x. **11.** ________________

$\log_7 3 - \frac{1}{2}\log_7 x = \log_7 \frac{1}{2}$

12. Simplify the expression. **12.** ________________

$\frac{3e^x}{e^{3x}} \cdot \frac{e^{-2x}}{12}$

13. Evaluate to three decimal places. **13.** ________________

$e^{1.35}$

14. Is $f(x) = 7e^{-2t}$ an example of exponential growth or decay? **14.** ________________

15. Sketch the graph of the function. **15.** *Use graph at left.*

$f(x) = \frac{1}{3}e^x - 2$

Name ____________________

16. Expand the expression. $\ln \frac{3xy^2}{19}$

16. ____________________

17. Condense the expression. $\frac{1}{3}\ln x - 2\ln y$

17. ____________________

18. Evaluate $5\ln(7.4)$.

18. ____________________

19. Evaluate $\log_7 25$ to three decimal places.

19. ____________________

20. Sketch the graph of the function.

$$f(x) = \ln(x + 3)$$

20. *Use graph at left.*

21. Solve the equation. $35\log_6 x = 18$

21. ____________________

22. ***Double the Balance*** How long does it take \$100 to double if invested at 6% interest compounded continuously?

22. ____________________

23. Sketch the graph of the function. Identify the horizontal asymptotes.

$$f(x) = \frac{5}{1 + 4e^{-x}}$$

23. ____________________
Use graph at left.

Chapter 8 Test

Form B

(Page 1 of 3 pages)

Name ______________________

Date ______________________

1. Evaluate $5^{\sqrt{3}/2}$ to three decimal places. 1. ______________

2. ***Company Sales*** A company had total sales of \$3,200,000 in 1985. Each year between 1985 and 1991 the sales increased by 12%. Approximate the sales for 1991 to the nearest \$100,000. 2. ______________

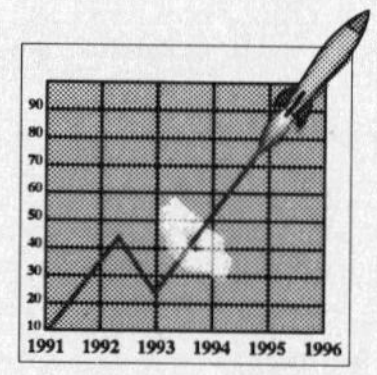

3. Sketch the graph.

   $f(x) = 2^{x-1}$

3. *Use graph at left.*

4. Write $27^{2/3} = 9$ in logarithmic form. 4. ______________

5. Evaluate without using a calculator. 5. ______________

   $\log_7 \frac{1}{49}$

6. Which of the following is *undefined*? 6. ______________

   $\log_3 9,\ \log_3 1,\ \log_3(-3)$

7. Evaluate $\log_6 30$ to three decimal places. 7. ______________

Name ____________________

8. Solve for x. $\log_5 \frac{1}{125} = x$ **8.** ____________

9. Expand the expression. $\log_4 3x^{-1}y^3$ **9.** ____________

10. Condense the expression. **10.** ____________

$\frac{1}{3}\log_5 8 - \frac{1}{2}\log_5 x + 4\log_5 y$

11. Solve for x. **11.** ____________

$\log_{10} 8 - \frac{1}{3}\log_{10} x = \log_{10} 2$

12. Simplify the expression. **12.** ____________

$\frac{5e^{-2x}}{e^{4x}} \cdot \frac{(e^{2x})^2}{10e^{-x}}$

13. Evaluate $e^{2.53}$ to three decimal places. **13.** ____________

14. Is $f(x) = 10.2e^{0.04t}$ an example of exponential growth or decay? **14.** ____________

15. Sketch the graph of the function.

$f(x) = 2e^x - 1$

15. *Use graph at left.*

Name ______________________

16. Expand the expression. $\ln \dfrac{5y^2x^3}{4}$ — **16.** ______________

17. Condense the expression. $\frac{1}{2}\ln x - \frac{3}{2}\ln y$ — **17.** ______________

18. Evaluate $14\ln(17.8)$ to three decimal places. — **18.** ______________

19. Evaluate $\log_9 125$ to three decimal places. — **19.** ______________

20. Sketch the graph of the function.

$$f(x) = \ln(x-1)$$

20. *Use graph at left.*

21. Solve the equation. $75\log_7 x = 61$ — **21.** ______________

22. ***Double the Balance*** How long does it take \$1000 to double if invested at 7% interest compounded continuously? — **22.** ______________

23. Sketch the graph of the function. Identify the horizontal asymptotes.

$$f(x) = \frac{4}{2+e^{-x}}$$

23. ______________ *Use graph at left.*

Chapter 8 Test

Form C

(Page 1 of 3 pages)

Name ______________________

Date ______________________

1. Evaluate π^{-2} to two decimal places.

1. ______________________

2. ***Company Sales*** A company had a total debt of \$320,000 in 1980. Between 1980 and 1987 it was able to reduce its debt 15% each year. Approximate the company's debt in 1987 to the nearest \$1000.

2. ______________________

3. Sketch the graph.

   $f(x) = 2 - \left(\frac{1}{2}\right)^x$

3. *Use graph at left.*

4. Write $64^{-2/3} = \frac{1}{16}$ in logarithmic form.

4. ______________________

5. Evaluate without using a calculator.

   $\log_7 7^{\sqrt{2}}$

5. ______________________

6. Which of the following is *undefined*?

   $\log_4 \frac{1}{2},\ \log_4(-6),\ \log_4 1$

6. ______________________

7. Evaluate $\log_{12} 782$ to three decimal places.

7. ______________________

 Name ________________

8. Solve for x. $\log_x \frac{8}{27} = -3$ **8.** ________________

9. Expand the expression. $\log_5 \left(5\sqrt{\frac{x}{y}}\right)$ **9.** ________________

10. Condense the expression. **10.** ________________

$\frac{1}{5}\log_3 32 - 2\log_3 x + \frac{1}{2}\log_3 y$

11. Solve for x. **11.** ________________

$2\log_5 10 - \frac{1}{4}\log_5 x = \log_5 50$

12. Simplify the expression. **12.** ________________

$$\frac{5(e^{2x})^{-1}}{2e^{-3x}} \cdot \frac{(5e^{x})^{2}}{(e^{-x})^{-1}}$$

13. Evaluate $e^{2\pi}$ to three decimal places. **13.** ________________

14. Is $f(x) = 13.7e^{-0.04t}$ an example of exponential growth or decay? **14.** ________________

15. Sketch the graph of the function. **15.** *Use graph at left.*

$f(x) = 2 - e^{-x}$

Name ______________________

16. Condense the expression. $\frac{2}{3}\ln x - \frac{1}{3}\ln 2y$ **16.** ______________

17. Evaluate $\dfrac{\ln[(3.5)^4]}{2.17}$ to three decimal places. **17.** ______________

18. Evaluate $\log_9 152$ to three decimal places. **18.** ______________

19. Sketch the graph of the function. **19.** *Use graph at left.*

$f(x) = \ln(3 - x)$

20. Solve the equation. $7\log_5 x - 3 = 15$ **20.** ______________

21. Solve the equation. $6^{-0.2x} - 3 = 7$ **21.** ______________

22. ***Double the Balance*** How long does it take \$1000 to double if invested at 6% annual interest compounded quarterly? **22.** ______________

23. Sketch the graph of the function. Identify the horizontal asymptotes. **23.** ______________ *Use graph at left.*

$$f(x) = \frac{6}{2 + e^{-x}}$$

Mid-Chapter 9 Test Form A

Name ____________________

Date ____________________

1. Perform the indicated operation.

$(5x^3 - 2x + 3) - (-2x^3 + x^2 - 4)$

1. ____________________

2. Perform the indicated operations.

$(x + 4)(x^2 - 5x + 2)$

2. ____________________

3. Sketch the graph of the function.

$f(x) = -(x - 1)^3$

3. *Use graph at left.*

4. Factor completely with respect to the integers.

$128x^3 - 2$

4. ____________________

5. Find all real-number solutions.

$x^3 + 2x^2 - 9x - 18 = 0$

5. ____________________

6. Use long division. $(x^3 - 3x^2 + 2x - 8) \div (x^2 + 2)$

6. ____________________

7. Use synthetic division. $(3x^2 - 5x + 7) \div (x - 3)$

7. ____________________

8. Use the Remainder Theorem to evaluate $f(x)$ at $x = 3$.

$f(x) = 4x^4 - 8x^3 - 9x^2 - 12x + 5$

8. ____________________

Mid-Chapter 9 *Test* *Form B*

Name ______________________

Date ______________________

1. Perform the indicated operation. — 1. ____________

   $(7x^3 + 5x^2 - 7) - (-x^3 + 3x - 5)$

2. Perform the indicated operations. — 2. ____________

   $(x + 3)(x^2 - 4x + 5)$

3. Sketch the graph of the function. — 3. *Use graph at left.*

   $f(x) = 2 - \frac{x^4}{8}$

4. Factor completely with respect to the integers. — 4. ____________

   $81x^3 + 24$

5. Find all real-number solutions. — 5. ____________

   $x^4 - x^2 - 12 = 0$

6. Use long division. $(x^3 + 2x^2 + 4x + 18) \div (x^2 + 4)$ — 6. ____________

7. Use synthetic division. $(5x^2 - 2x - 7) \div (x + 2)$ — 7. ____________

8. Use the Remainder Theorem to evaluate $f(x)$ at $x = -3$. — 8. ____________

   $f(x) = 4x^3 + 8x^2 - 9x + 12$

Chapter 9 Test

Form A

(Page 1 of 3 pages)

Name ______________________

Date ______________________

1. Perform the indicated operation. **1.** ____________

$(2x + 5) - (3x^2 + 7x - 5)$

2. Perform the indicated operations. **2.** ____________

$(3x + 2)(2x^2 - 7x - 4)$

3. State the maximum number of turns in the graph of **3.** ____________

$f(x) = 2x^3 - 2x^2 + 3.$

4. State the left and right behaviors of the graph of **4.** ____________

$f(x) = -x^3 + 7x + 4.$

5. Sketch the graph of the function. **5.** *Use graph at left.*

$f(x) = (x + 1)^4$

6. Factor completely with respect to the integers. **6.** ____________

$10x^4 - 160$

7. Factor completely with respect to the integers. **7.** ____________

$4x^3 - 8x^2 + 3x - 6$

8. Find all real-number solutions. **8.** ____________

$x^3 + 6x^2 + 12x + 8 = 0$

 Name ____________

9. Use long division. Write the result in fractional form. **9.** ____________

$(5x^4 - 3x^2 + 4) \div (x^2 + 2)$

10. Use synthetic division. Write the result in fractional form. **10.** ____________

$(2x^3 + 9x^2 + 3x - 6) \div (x + 4)$

11. Use the Remainder Theorem to evaluate the function. **11.** ____________

$f(x) = 2x^4 - 12x^2 - 20x - 3$ at $x = 3$

12. Write a polynomial function whose graph has the given x-intercepts and has a leading coefficient of 1. **12.** ____________

$(2,\ 0), (-2,\ 0), (1,\ 0)$

13. Write the polynomial as a product of linear factors. **13.** ____________

$x^3 - 2x^2 - 9x + 18$

14. Find all real zeros of the function. **14.** ____________

$g(x) = 2x^3 - x^2 - 10x + 5$

15. ***Tree Heights*** The heights (in inches) of ten 5-year old maple trees are given below. Find the *range*, the *mean*, and the *standard deviation* of the data. **15.** ____________

53, 47, 44, 47, 51, 45, 46, 47, 46, 50

16. ***Geometry*** Write an expression for the area inside the rectangle but outside the square. **16.** ____________

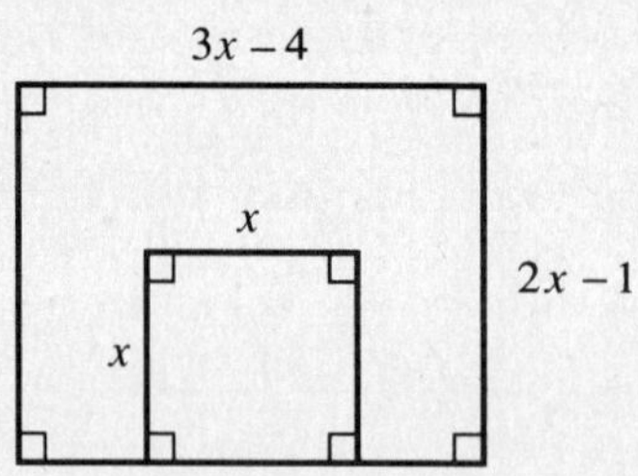

Chapter 9 Test

Form B

Name ______________________

Date ______________________

1. Perform the indicated operation.

$(3x + 7) - (5x^2 - 6x + 2)$

1. ______________________

2. Perform the indicated operations.

$(4x - 1)(3x^2 - 2x - 4)$

2. ______________________

3. State the maximum number of turns in the graph of

$f(x) = x^4 - x^2 + 3x - 7.$

3. ______________________

4. State the left and right behaviors of the graph of

$f(x) = 3x^3 - x^2 + 4.$

4. ______________________

5. Sketch the graph of the function.

$f(x) = (x + 1)^3$

5. *Use graph at left.*

6. Factor completely with respect to the integers.

$5x^4 + 10x^2 - 15$

6. ______________________

7. Factor completely with respect to the integers.

$2x^4 - 14x^2 + 24$

7. ______________________

8. Find all real-number solutions.

$x^3 - 9x^2 + 27x - 27 = 0$

8. ______________________

9. Use long division. Write the result in fractional form. 9. ____________

$(6x^4 + 20x^2 - 4) \div (x^2 + 3)$

10. Use synthetic division. Write the result in fractional form. 10. ____________

$(2x^3 + 6x^2 - 16x + 24) \div (x + 5)$

11. Use the Remainder Theorem to evaluate the function. 11. ____________

$f(x) = 3x^3 + 10x^2 - 2x - 21$ at $x = -3$

12. Write a polynomial function whose graph has the given x-intercepts and has a leading coefficient of 1. 12. ____________

$(-3,\ 0), (0,\ 0), (5,\ 0)$

 Name ______________________

13. Write the polynomial as a product of linear factors. **13.** ______________

$x^3 + 3x^2 - 4x - 12$

14. Find all real zeros of the function. **14.** ______________

$g(x) = 2x^3 + x^2 - 6x - 3$

15. ***Tree Height*** The heights (in inches) of ten 3-year old maple trees are given below. Find the *range*, the *mean*, and the *standard deviation* of the data. **15.** ______________

33, 35, 39, 34, 35, 37, 31, 35, 30, 29

16. ***Geometry*** Write an expression for the area inside the rectangle but outside the square. **16.** ______________

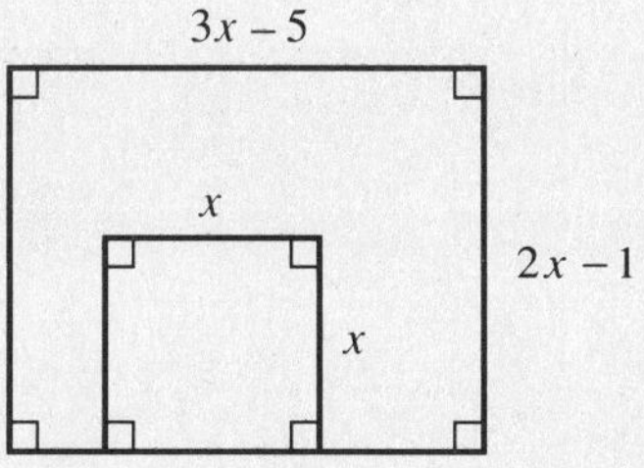

Name ____________________

Date ____________________

1. Perform the indicated operation. — 1. ________

   $(-7x^2 + 3) - (4x^2 + 2x - 1)$

2. Perform the indicated operations. — 2. ________

   $x(2x - 1)(x^2 + 3)$

3. State the maximum number of turns in the graph of — 3. ________

   $g(x) = 2x^5 - x^3 + 2x - 1$.

4. State the left and right behaviors of the graph of — 4. ________

   $f(x) = -x^3 + x^2 - 7$.

5. Sketch the graph of the function. — 5. *Use graph at left.*

   $f(x) = 4 - \frac{x^4}{4}$

6. Factor completely with respect to the integers. — 6. ________

   $x^4 - 13x^2 + 36$

7. Factor completely with respect to the integers. — 7. ________

   $2x^3 - 3x^2 + 4x - 6$

8. Find all real-number solutions. — 8. ________

   $x^3 - 5x^2 + 8x - 4 = 0$

Name ______________________

9. Use long division. Write the result in fractional form. **9.** ______________

$(4x^4 + 2x^3) \div (2x^2 - 3)$

10. Use synthetic division. Write the result in fractional form. **10.** ______________

$(2x^3 - 7x^2 - 6x + 15) \div (x - 4)$

11. Use the Remainder Theorem to evaluate the function. **11.** ______________

$f(x) = x^3 + 7x^2 + 6x - 10$ at $x = -5$

12. Write a polynomial function whose graph has the given x-intercepts and has a leading coefficient of 1. **12.** ______________

$(-1,\ 0), (\frac{1}{3},\ 0), (\frac{1}{2},\ 0)$

 Name ____________

13. Write the polynomial as a product of linear factors. **13.** ____________

$4x^3 - 8x^2 - x + 2$

14. Find all real zeros of the function. **14.** ____________

$g(x) = x^3 - 2x + 1$

15. ***Tree Heights*** The heights (in inches) of ten 6-year old maple trees are given below. Find the *range*, the *mean*, and the *standard deviation* of the data. **15.** ____________

64.2, 71.3, 68.4, 69.5, 68.3, 70.9, 67.2, 70.4, 71.9, 66.1

16. ***Geometry*** Write an expression for the area inside the rectangle but outside the triangle. **16.** ____________

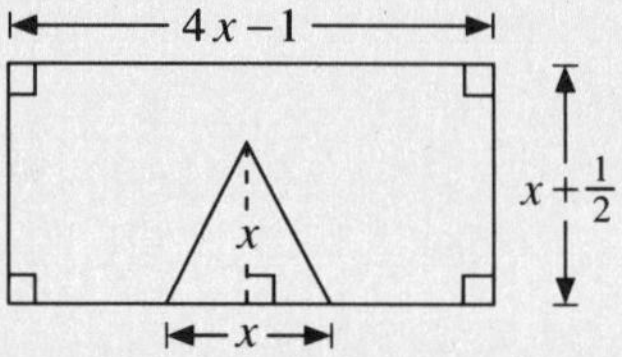

Cumulative Test

Chapters 7–9

(Page 1 of 8 pages)

Name ______________________

Date ______________________

1. Simplify $(-4x^2)^{-3}$. 1. ______________

2. Simplify $\dfrac{3x^{-2}y^3}{(2y)^2} \cdot \dfrac{(2x^2y^{-2})^2}{3^{-2}}$. 2. ______________

3. Solve for x. $4^3 \cdot 4^{x-1} \cdot 4^5 = 4^{2x}$ 3. ______________

4. ***Depreciated Value*** A \$20,000 automobile depreciates 17% each year for the first five years. Find its value, to the nearest \$1000, after the first five years. 4. ______________

5. ***Balance in an Account*** \$1500 is deposited in an account paying 6.5% annual interest, compounded quarterly. Find its value after 5 years. 5. ______________

6. Evaluate $(64)^{5/6}$. 6. ______________

7. Evaluate $(12)^{2/3}$ to three decimal places. **7.** ______________

8. ***Geometry*** The volume of a sphere is $\frac{4}{3}\pi r^3$, where r is the radius. Find, to two decimal places, the radius of a sphere whose volume is 200 cubic inches. **8.** ______________

9. Evaluate $\sqrt[7]{10194}$ to three decimal places. **9.** ______________

10. Simplify $49^{2/3} \cdot 49^{-1/6}$. **10.** ______________

11. Simplify $\sqrt[3]{16} + \sqrt[3]{54} - \sqrt[3]{2}$. **11.** ______________

12. Solve the equation $\sqrt[4]{3y - 5} = 2$. **12.** ______________

Name ______________________

13. Find the distance between the points.

$(2,\ -7)$ and $(-3,\ 5)$

13. ______________

14. Find the geometric mean of 7 and 27. Round your result to two decimal places.

14. ______________

15. State the *domain* and *range* of $g(x)$ and sketch its graph.

$g(x) = 2 - \sqrt{x+1}$

15. ______________
Use graph at left.

16. Expand the expression $\ln \sqrt[3]{\dfrac{xy^2}{27}}$.

16. ______________

17. Solve for x. $\log_5 \frac{1}{125} = x$

17. ______________

Name ____________________

18. Condense the expression.

$\frac{1}{2}[5 \log_3 x - 2 \log_3 y]$

18. ____________________

19. Evaluate $e^{3.5}$ to two decimal places.

19. ____________________

20. Is $f(x) = 6e^{0.2t}$ an example of exponential growth or decay?

20. ____________________

21. Solve for x. $\log_6 4 - \frac{1}{2} \log_6 x = \log_6 \frac{1}{3}$

21. ____________________

22. Sketch the graph of the function.

$f(x) = 2e^{-x/2}$

22. *Use graph at left.*

23. Evaluate $\ln \sqrt{257}$ to three decimal places.

23. ____________________

Name ______________________

24. Evaluate $\log_5 53$ to three decimal places.

24. ______________________

25. ***Double the Balance*** How long does it take \$100 to double if invested at 5.5% annual interest compounded continuously?

25. ______________________

26. Sketch the graph of the function.

$$f(x) = \ln(x - 1)$$

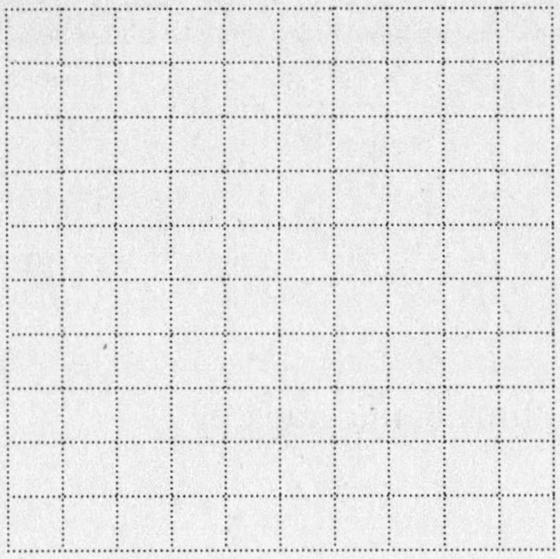

26. *Use graph at left.*

27. Sketch the graph of the function. Identify the horizontal asymptotes.

$$g(x) = \frac{3}{1 + e^{-x}}$$

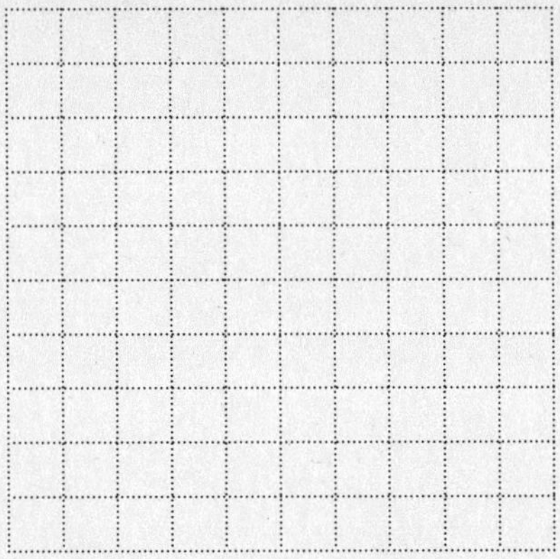

27. ______________________ *Use graph at left.*

 Name ______________________

28. Perform the indicated operations. **28.** ______________

$(2x - 3)(x^2 + 4) - x(3x - 5)$

29. Sketch the graph of the function. **29.** *Use graph at left.*

$f(x) = -(x + 1)^3$

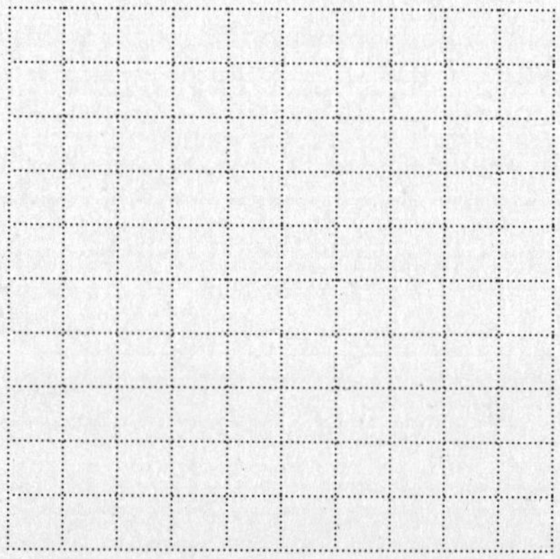

30. State the maximum number of turns in the graph of $f(x) = -x^4 - x^3 + 2.$ **30.** ______________

31. Factor completely with respect to the integers. **31.** ______________

$3x^3 - 81$

32. Factor completely with respect to the integers. **32.** ______

$2x^3 - 3x^2 + 6x - 9$

33. Find all real-number solutions. **33.** ______

$x^3 - 3x - 2 = 0$

34. Use long division. Write the answer in fractional form. **34.** ______

$(2x^4 + x^3 - 8x^2) \div (x^2 - 4)$

35. Use the Remainder Theorem to evaluate the function. **35.** ______

$f(x) = 2x^3 + 15x^2 + 8x + 5$ at $x = -7$

36. Write a polynomial whose graph has the given x-intercepts and has a leading coefficient of 1. **36.** ______

$(3,\ 0),\ (-3,\ 0),\ (2,\ 0)$

37. One root of the equation is $x = \frac{1}{4}$. Find the other roots. **37.** ____________

$8x^3 - 30x^2 + 19x - 3 = 0$

38. Find all real zeros of the function. **38.** ____________

$g(x) = x^3 - 3x^2 - 3x + 9$

39. Write the polynomial as a product of linear factors. **39.** ____________

$x^3 - 3x^2 - 10x + 24$

40. ***Weights*** The weights (in pounds) of ten 3-year old basset hounds are given below. Find the *range*, the *mean*, and the *standard deviation* of the data. **40.** ____________

33.1, 27.4, 37.8, 29.2, 30.1, 40.3, 35.2, 31.4, 32.1, 33.4

Mid-Chapter 10 Test Form A

Name ____________________

Date ____________________

1. Identify all horizontal and vertical asymptotes of the graph of the function.

$$f(x) = \frac{x+1}{x(x-3)}$$

1. ____________________

2. Sketch the graph of the function.

$$f(x) = \frac{x^2+3}{3-x^2}$$

2. *Use graph at left.*

3. ***Typing Time*** The time, t (in minutes), you need to type a term paper varies inversely with the rate, r (in words per minute), that you can type. When typing 50 words per minute, you can finish in 80 minutes. How long would it take if you could type only 40 words per minute?

3. ____________________

4. z varies jointly as the product of x and y. Find the equation that relates the variables if $z = 5.6$ when $x = 8$ and $y = 0.028$.

4. ____________________

5. Divide and simplify.

$$\frac{x^2-x-12}{x^2+3x} \div \frac{2x-8}{x^2-9}$$

5. ____________________

6. Perform the operations and simplify.

$$\frac{x^2+5x}{x^2-1} \cdot \frac{x^2-x}{x^2+7x+10} \cdot \frac{(x+1)^2}{x^3+x^2}$$

6. ____________________

Mid-Chapter 10 Test Form B

Name ______________________

Date ______________________

1. Identify all horizontal and vertical asymptotes of the graph of the function.

$$f(x) = \frac{x+5}{x(x-2)}$$

1. ______________________

2. Sketch the graph of the function.

$$f(x) = \frac{x}{x^2-4}$$

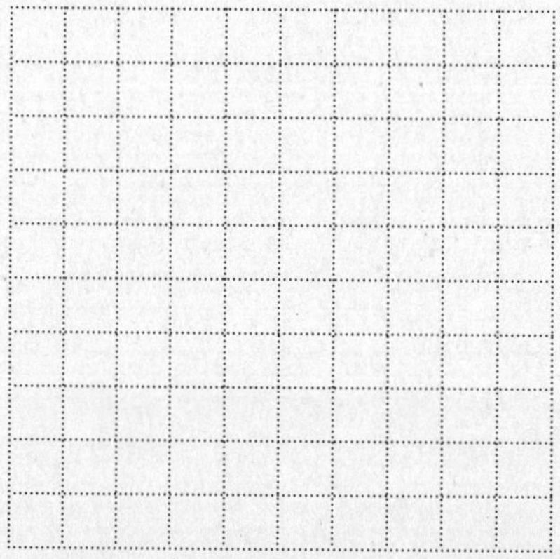

2. *Use graph at left.*

3. ***Typing Time*** The time, t (in minutes), you need to type a term paper varies inversely with the rate, r (in words per minute), that you can type. When typing 60 words per minute, you can finish in 80 minutes. How long would it take if you could type only 40 words per minute?

3. ______________________

4. z varies jointly as the product of x and y. Find the equation that relates the variables if $z = 175$ when $x = 7.5$ and $y = \frac{10}{3}$.

4. ______________________

5. Divide and simplify.

$$\frac{x^2-x-6}{x^2-3x} \div \frac{x^2-4}{3x}$$

5. ______________________

6. Perform the operations and simplify.

$$\frac{4x^2-9}{x^2-x-2} \cdot \frac{x^2+x}{x^2} \cdot \frac{x^3-2x^2}{2x^2+3x}$$

6. ______________________

Name ____________________

Date ____________________

1. Identify all horizontal and vertical asymptotes of the graph of the function.

$$f(x) = \frac{x^2}{x^2 - 4}$$

1. ____________________

2. Sketch the graph of the function.

$$f(x) = \frac{x - 2}{x + 2}$$

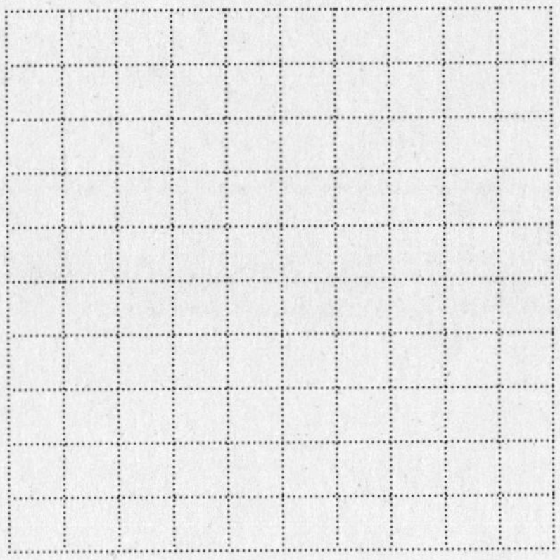

2. *Use graph at left.*

3. x and y vary inversely. $x = 7$ when $y = -4$. Find an equation that relates the variables.

3. ____________________

4. z varies jointly with the product of x and y. $z = 2.4$ when $x = 3$ and $y = 2$. Find an equation that relates the variables.

4. ____________________

5. Simplify the expression.

$$\frac{x^2 - 2x - 3}{x^2 - 1}$$

5. ____________________

6. Multiply and simplify.

$$\frac{(x + 2)^2}{x - 5} \cdot \frac{x^2 - 2x}{x^2 - 4}$$

6. ____________________

 Name ______________

7. Divide and simplify. 7. ______________

$$\frac{x^2+8x-20}{5x^3+50x^2} \div \frac{x^2+9x}{x^2+7x-18}$$

8. Simplify. $\frac{(2x^2y^3)^2}{(x^3y^2)^3} \div \frac{(4x)^2y^3}{(xy)^4}$ 8. ______________

9. Find the least common multiple. 9. ______________

$x^3,\ x^2,\ 3,\ (x-2),\ x^2-4$

10. Is $x = -3$ a solution of 10. ______________

$$\frac{x+4}{x+3} = 2 + \frac{1}{x+3}?$$

11. Find all of the zeros of the function. 11. ______________

$$f(x) = \frac{x^2-3x-40}{x^2+x+1}$$

12. Solve the equation. 12. ______________

$$\frac{x}{30} - \frac{1}{5x} = \frac{1}{6}$$

 Name ____________

13. ***Average Cost*** Startup costs for producing a product are \$10,000. Thereafter, each item costs \$5 to produce. How many must be produced to bring the *average cost* per item down to \$10? **13.** ____________

14. Perform the operations and simplify. **14.** ____________

$$\frac{3x+4}{x^2-16}-\frac{2}{x-4}$$

15. Simplify the complex fraction. **15.** ____________

$$\frac{\dfrac{5}{x+2}}{3-\dfrac{2}{x+2}}$$

16. Solve the equation. **16.** ____________

$$\frac{\dfrac{x-3}{2}+\dfrac{1}{6}}{\dfrac{1}{12}-\dfrac{x-2}{3}}=-1$$

 Name ____________________

Home Mortgage **The table gives the *monthly payment* on a home mortgage of $100,000 for several times and interest rates.**

	9%	10%	11%
15 years	1074.61	1136.60	1200.17
20 years	965.02	1032.19	1101.09
25 years	908.70	980.11	1053.72
30 years	877.57	952.32	1028.61

17. Use the table above to find the total interest payment on a $100,000 home mortgage at 10% interest for 20 years. **17.** ____________

18. In Problem 17, how much money would have been saved if the mortgage had been for 15 years instead of 20? **18.** ____________

Installment Loan **The monthly payment, *M* (in dollars), on an installment loan of principal, *P* (in dollars), at an annual rate, *r*, for *t* years is given by the formula**

$$M = P\left[\frac{i}{1 - \left(\frac{1}{1+i}\right)^{12t}}\right]$$

where $i = \frac{1}{12}r$. Use the formula in Problem 19 and 20.

19. You borrow $10,000 at an annual interest rate of 12% to be repaid in 5 years. Find the monthly payment. **19.** ____________

20. Find the total interest payment of the loan in Problem 19. **20.** ____________

Chapter 10 Test Form B

(Page 1 of 4 pages)

Name ____________________

Date ____________________

1. Identify all horizontal and vertical asymptotes of the graph of the function.

$$f(x) = \frac{5x}{x^2 - 1}$$

1. ____________________

2. Sketch the graph of the function.

$$f(x) = \frac{4 - 2x}{x - 3}$$

2. *Use graph at left.*

3. x and y vary inversely. $x = -4$ when $y = 18$. Find an equation that relates the variables.

3. ____________________

4. z varies jointly with the product of x and y. $z = 3.6$ when $x = 8$ and $y = 90$. Find an equation that relates the variables.

4. ____________________

5. Simplify the expression.

$$\frac{x^3 + 4x^2}{x^2 - 16}$$

5. ____________________

6. Multiply and simplify.

$$\frac{2x - 3}{(x + 3)^2} \cdot \frac{x^2 + 4x + 3}{4x^2 - 9}$$

6. ____________________

 Name ____________

7. Divide and simplify. 7. ____________

$$\frac{(x+7)}{x^2+5x-14} \div \frac{x^2+x-6}{x+3}$$

8. Simplify. $\frac{(xy^2)^2}{(2xy)^3} \div \frac{x^2y^4}{(4x)^2y^3}$ 8. ____________

9. Find the least common multiple. 9. ____________

$2x^2,\ 4x,\ x^3,\ (x-1),\ (x-1)^2$

10. Is $x = -2$ a solution of 10. ____________

$$\frac{4x}{x+2} = 1 - \frac{8}{x+2}?$$

11. Find all of the zeros of the function. 11. ____________

$$f(x) = \frac{x^2-4x-12}{(x+3)^2}$$

12. Solve the equation. 12. ____________

$$\frac{x}{2} + \frac{1}{15} = \frac{1}{30x}$$

 Name ____________

13. ***Average Cost*** Startup costs for producing a product are \$12,000. Thereafter, each item costs \$3 to produce. How many must be produced to bring the *average cost* per item down to \$6? **13.** ____________

14. Perform the operations and simplify. **14.** ____________

$$\frac{3x-5}{x^2-25}-\frac{2}{x+5}$$

15. Simplify the complex fraction. **15.** ____________

$$\frac{\frac{4}{x-3}+\frac{2}{3}}{\frac{5}{x-3}}$$

16. Solve the equation. **16.** ____________

$$\frac{\frac{x+1}{4}-\frac{1}{2}}{\frac{1}{8}+\frac{x+2}{4}}=-2$$

Name ______________________

Home Mortgage **The table gives the *monthly payment* on a home mortgage of $100,000 for several times and interest rates.**

	9%	10%	11%
15 years	1074.61	1136.60	1200.17
20 years	965.02	1032.19	1101.09
25 years	908.70	980.11	1053.72
30 years	877.57	952.32	1028.61

17. Use the table above to find the total interest payment on a $100,000 home mortgage at 9% interest for 25 years. **17.** ____________

18. In Problem 17, how much money would have been saved if the mortgage had been for 20 years instead of 25? **18.** ____________

Installment Loan **The monthly payment, *M* (in dollars), on an installment loan of principal, *P* (in dollars), at an annual rate, *r*, for *t* years is given by the formula**

$$M = P\left[\frac{i}{1 - \left(\frac{1}{1+i}\right)^{12t}}\right]$$

where $i = \frac{1}{12}r$. Use the formula in Problem 19 and 20.

19. You borrow $10,000 at an annual interest rate of 12% to be repaid in 8 years. Find the monthly payment. **19.** ____________

20. Find the total interest payment of the loan in Problem 19. **20.** ____________

Chapter 10 Test Form C

Name ______________________

Date ______________________

1. Identify all horizontal and vertical asymptotes of the graph of the function.

$$f(x) = \frac{-x^3}{x^3 - 8}$$

1. ______________________

2. Sketch the graph of the function.

$$f(x) = \frac{x^2}{x^2 - 9}$$

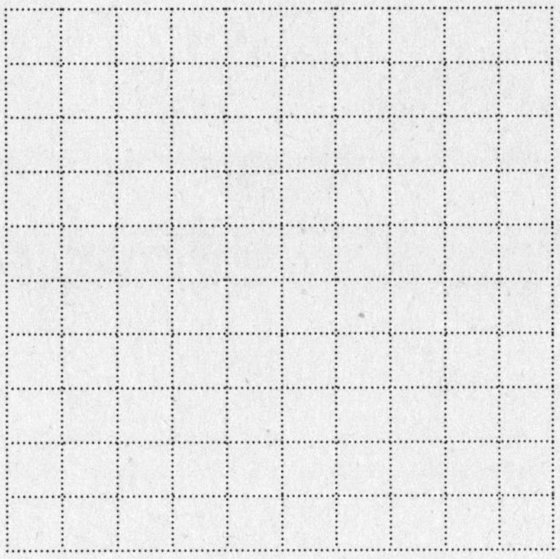

2. *Use graph at left.*

3. x and y vary inversely. $y = \frac{3}{5}$ when $x = \frac{7}{2}$. Find an equation that relates the variables.

3. ______________________

4. z varies jointly with the product of x and y. $z = 1.44$ when $x = 3.6$ and $y = 8$. Find an equation that relates the variables.

4. ______________________

5. Simplify the expression.

$$\frac{x^2 - x - 12}{x^2 + x - 20}$$

5. ______________________

6. Multiply and simplify.

$$\frac{x^2 + 4x}{x^2 - 6x + 8} \cdot \frac{x^2 - x - 2}{3x^3 + 12x^2}$$

6. ______________________

Name ______________________

7. Perform the indicated operations and simplify. **7.** ______________

$$\frac{x+2}{x+9} \cdot \left[\frac{x^2+9x}{x^2-4} \div \frac{3x^2+6x}{x^2+2x}\right]$$

8. Simplify. $\frac{x^2y^3}{3x^4} \cdot \frac{(xy)^2}{x^3y} \div \frac{x^2y^4}{6y^3}$ **8.** ______________

9. Find the least common multiple. **9.** ______________

$9x,\ 3x^2,\ (x^2-9),\ (x-3)^2$

10. Is $x = 2$ a solution of **10.** ______________

$$\frac{4x}{x+2} = 4 - \frac{8}{x+2}?$$

11. Find all of the zeros of the function. **11.** ______________

$$f(x) = \frac{x^2(x-2)^2}{(x+2)^2}$$

12. Solve the equation. **12.** ______________

$$\frac{2x}{x-2} = \frac{1}{x^2-4} + 1$$

Name ______________________

13. ***Average Cost*** Startup costs for producing a product are \$15,000. Thereafter, each item costs \$7 to produce. How many must be produced to bring the *average cost* per item down to \$9.50?

13. ______________________

14. Perform the operations and simplify.

$$\frac{4x}{x^2 - 9} + \frac{2}{x + 3} - \frac{2}{x - 3}$$

14. ______________________

15. Simplify the complex fraction.

$$\frac{\dfrac{2}{x+2} - \dfrac{3}{x}}{\dfrac{3}{x+2} + \dfrac{2}{x}}$$

15. ______________________

16. Solve the equation.

$$\frac{\dfrac{2x-3}{2} + \dfrac{2}{3}}{\dfrac{1}{2} - \dfrac{x+1}{3}} = 4$$

16. ______________________

Name ____________________

Home Mortgage **The table gives the *monthly payment* on a home mortgage of $100,000 for several times and interest rates.**

	9%	10%	11%
15 years	1074.61	1136.60	1200.17
20 years	965.02	1032.19	1101.09
25 years	908.70	980.11	1053.72
30 years	877.57	952.32	1028.61

17. Use the table above to compare the total interest payment on a $100,000 home mortgage at 11% interest for 15 years and 20 years. **17.** ____________

18. In Problem 17, how much money would have been saved if a 15-year mortgage were at 10% instead of at 11%? **18.** ____________

Installment Loan **The monthly payment, *M* (in dollars), on an installment loan of principal, *P* (in dollars), at an annual rate, *r*, for *t* years is given by the formula**

$$M = P\left[\frac{i}{1 - \left(\frac{1}{1+i}\right)^{12t}}\right]$$

where $i = \frac{1}{12}r$. Use the formula in Problem 19 and 20.

19. You borrow $10,000 at an annual interest rate of 9.5% to be repaid in 4 years. Find the monthly payment. **19.** ____________

20. Find the total interest payment of the loan in Problem 19. **20.** ____________

Mid-Chapter 11 Test Form A

Name ______________________

Date ______________________

In 1–3, refer to the parabola $x^2 + 16y = 0$.

1. Find the *vertex* and *axis*. **1.** ____________

2. Find the *focus* and *directrix*. **2.** ____________

3. Sketch the graph of the parabola. **3.** *Use graph at left.*

4. A circle is centered at (0, 0) and passes through the point (−2, 5). Write its equation. **4.** ____________

5. Sketch the graph of

$x^2 + y^2 = 16.$

5. *Use graph at left.*

6. Find the foci and vertices of the ellipse. $\dfrac{x^2}{16} + \dfrac{y^2}{8} = 1$ **6.** ____________

7. Find an equation of the ellipse centered at (0, 0), with a vertex at (0, 6) and a co-vertex at (4, 0). Also find the *eccentricity*. **7.** ____________

8. Find the points of intersection, if any, of the graphs. **8.** ____________

$x^2 + y^2 = 5$

$x - y = 1$

Mid-Chapter 11 Test Form B

Name ______________________

Date ______________________

In 1–3, refer to the parabola $y^2 + 12x = 0$.

1. Find the *vertex* and *axis*. **1.** ______________

2. Find the *focus* and *directrix*. **2.** ______________

3. Sketch the graph of the parabola.

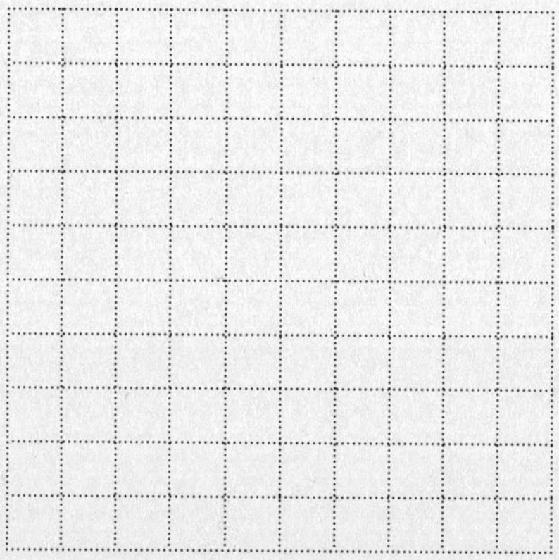

3. *Use graph at left.*

4. A circle is centered at (0, 0) and passes through the point (2, −3). Write its equation. **4.** ______________

5. Sketch the graph of

$x^2 + y^2 = 25.$

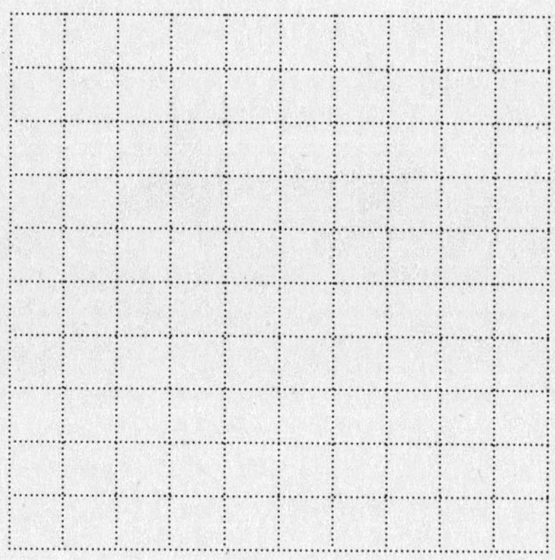

5. *Use graph at left.*

6. Find the foci and vertices of the ellipse. $\frac{x^2}{4} + \frac{y^2}{20} = 1$ **6.** ______________

7. Find an equation of the ellipse centered at (0, 0), with a vertex at (6, 0) and a co-vertex at (0, 3). Also find the *eccentricity*. **7.** ______________

8. Find the points of intersection, if any, of the graphs. **8.** ______________

$$x^2 = y$$

$$x - y = -2$$

Chapter 11 Test Form A

Name ____________________

Date ____________________

1. A parabola has vertex at (0, 0) and focus at (0, 5). Write its equation. 1. ____________

2. Find the *axis* and *directrix* of the parabola. 2. ____________

   $y^2 = -4x$

3. Sketch the graph of the parabola. 3. *Use graph at left.*

   $x^2 - 12y = 0$

4. Write an equation of the circle centered at (0, 0) with radius 7. 4. ____________

5. Sketch the graph of $x^2 + y^2 = 9$. 5. *Use graph at left.*

6. Find the points of intersection, if any, of the graphs. 6. ____________

   $y^2 + 4x = 0$

   $x + y = 1$

 Name ______________________

7. Find an equation of the ellipse centered at (0, 0), with a vertex at (9, 0) and a co-vertex at (0, 5).

7. ______________________

8. Find the *vertices*, the *foci*, and the *eccentricity* of the ellipse.

$$\frac{x^2}{49} + \frac{y^2}{16} = 1$$

8. ______________________

9. Sketch the graph of

$$\frac{x^2}{4} + \frac{y^2}{16} = 1.$$

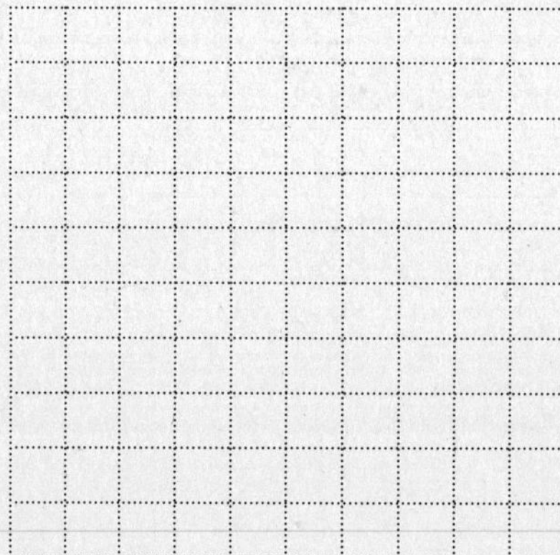

9. *Use graph at left.*

10. Find the *vertices* and the *foci* of the hyperbola.

$$\frac{y^2}{4} - \frac{x^2}{16} = 1$$

10. ______________________

11. A hyperbola has vertices at $(\pm 5,\ 0)$ and foci at $(\pm 7,\ 0)$. Write its equation.

11. ______________________

12. Find the asymptotes and sketch the hyperbola.

$$\frac{x^2}{1} - \frac{y^2}{4} = 1$$

12. ______________________
Use graph at left.

 Name ______________________

13. Identify the type of conic. If it is a circle, an ellipse, or a hyperbola, find its center. If it is a parabola, find its vertex. 13. ______________

$$\frac{(x-4)^2}{16}+\frac{(y+2)^2}{9}=1$$

14. Find an equation of the parabola with vertex at $(-3,\ 1)$ and focus $(-1,\ 1)$. 14. ______________

15. Write the equation of the circle in standard form. Identify the radius and center. 15. ______________

$$x^2+y^2-4x+6y+9=0$$

16. Classify the conic as a circle, an ellipse, a hyperbola, or a parabola. 16. ______________

$$5x^2-7y^2-3x+7y-14=0$$

Name ______________________

Date ______________________

1. A parabola has vertex at (0, 0) and focus at (−4, 0). Write its equation.

1. ______________________

2. Find the *axis* and *directrix* of the parabola.

   $x^2 = 12y$

2. ______________________

3. Sketch the graph of the parabola.

   $y^2 - 2x = 0$

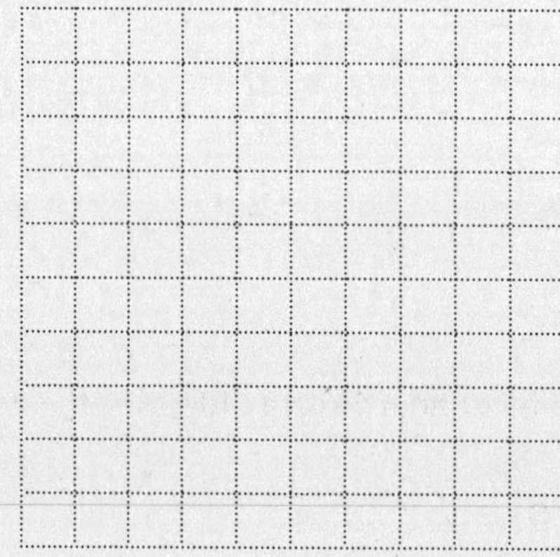

3. *Use graph at left.*

4. Write an equation of the circle centered at (0, 0) with radius 9.

4. ______________________

5. Sketch the graph of $x^2 + y^2 = 15$.

5. *Use graph at left.*

6. Find the points of intersection, if any, of the graphs.

   $y^2 + 4x = 0$

   $x + y = 0$

6. ______________________

 Name ____________________

7. Find an equation of the ellipse centered at (0, 0), with a vertex at (−8, 0) and a co-vertex at (0, 4). **7.** ____________

8. Find the *vertices*, the *foci*, and the *eccentricity* of the ellipse. **8.** ____________

$$\frac{x^2}{33} + \frac{y^2}{49} = 1$$

9. Sketch the graph of **9.** *Use graph at left.*

$$\frac{x^2}{25} + \frac{y^2}{4} = 1.$$

10. Find the *vertices* and the *foci* of the hyperbola. **10.** ____________

$$\frac{x^2}{16} - \frac{y^2}{4} = 1$$

11. A hyperbola has vertices at (±2, 0) and foci at (±8, 0). Write its equation. **11.** ____________

12. Find the asymptotes and sketch the hyperbola. **12.** ____________ *Use graph at left.*

$$\frac{y^2}{4} - \frac{x^2}{9} = 1$$

 Name ______________

13. Identify the type of conic. If it is a circle, an ellipse, or a hyperbola, find its center. If it is a parabola, find its vertex. **13.** ______________

$$\frac{(x+7)^2}{16} - \frac{(y-8)^2}{1} = 1$$

14. Find an equation of the ellipse with vertices at $(-2,\ 2)$ and $(4, 2)$, and co-vertices at $(1, 4)$ and $(1, 0)$. **14.** ______________

15. Write the equation of the circle in standard form. Identify the radius and center. **15.** ______________

$$x^2 + y^2 - 8x + 2y + 8 = 0$$

16. Identify the conic as a circle, an ellipse, a hyperbola, or a parabola. **16.** ______________

$$x^2 - 9y + 3x - 7 = 0$$

Chapter 11 Test Form C

(Page 1 of 3 pages)

Name ______________________

Date ______________________

1. Write an equation of the parabola with vertex at (0, 0) and directrix $y = -\frac{3}{2}$.

1. ______________

2. Find the *axis* and the *focus* of the parabola.

$$y^2 + 5x = 0$$

2. ______________

3. Sketch the graph of the parabola.

$$y^2 - 8x = 0$$

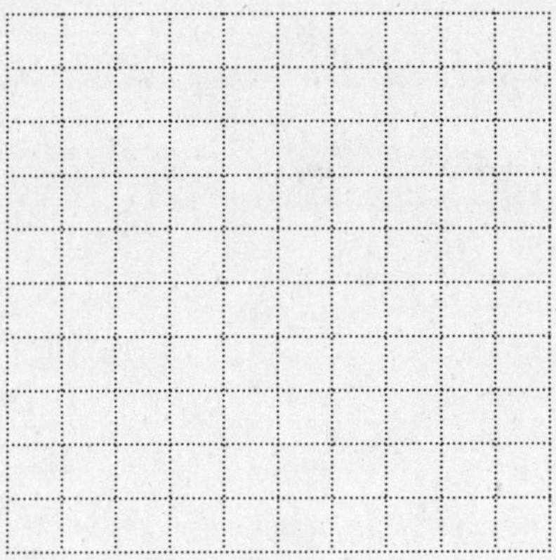

3. Use graph at left.

4. Write an equation of the circle centered at (0, 0) that passes through the point (1, −6).

4. ______________

5. Sketch the graph of

$$x^2 + y^2 = 10.$$

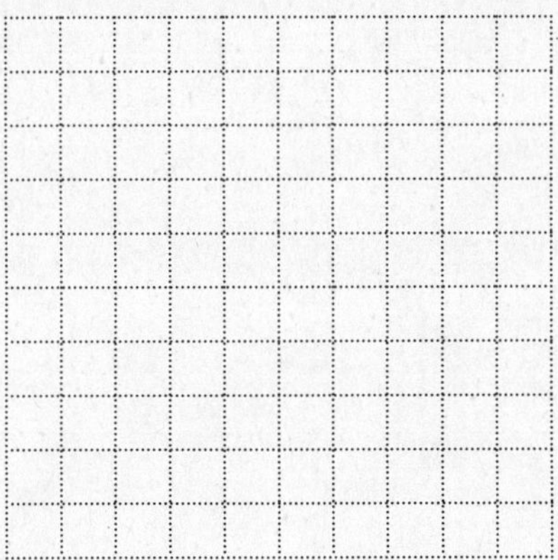

5. Use graph at left.

6. Find the points of intersection, if any, of the graphs.

$$x^2 + y^2 = 20$$
$$x + 2y = 0$$

6. ______________

 Name ______________________

7. Find an equation of the ellipse centered at (0, 0), with a vertex at (5, 0) and a focus at (4, 0). **7.** ______________

8. Find the *vertices*, the *foci*, and the *eccentricity* of the ellipse. **8.** ______________

$$\frac{x^2}{32} + \frac{y^2}{50} = 1$$

9. Sketch the graph of **9.** *Use graph at left.*

$$\frac{x^2}{4} + \frac{y^2}{9} = 1.$$

10. Find the *vertices* and the *foci* of the hyperbola. **10.** ______________

$$\frac{y^2}{5} - \frac{x^2}{11} = 1$$

11. A hyperbola has vertices at $(0, \pm 4)$ and foci at $(0, \pm 9)$. Write its equation. **11.** ______________

12. Find the asymptotes and sketch the hyperbola. **12.** ______________ *Use graph at left.*

$$x^2 - 4y^2 = 4$$

Name ______________________

13. Identify the type of conic. If it is a circle, an ellipse, or a hyperbola, find its center. If it is a parabola, find its vertex.

$(y+3)^2 = -27(x+7)$

13. ______________________

14. Find an equation of the hyperbola with vertices at $(-3, 2)$ and $(3, 2)$, and foci at $(-5, 2)$, $(5, 2)$.

14. ______________________

15. Write the equation of the circle in standard form. Identify the radius and center.

$x^2 + y^2 + 6x - 6y + 9 = 0$

15. ______________________

16. Classify the conic as a circle, an ellipse, a hyperbola, or a parabola.

$5y^2 + 4x^2 + 2y - 7x - 4 = 0$

16. ______________________

Mid-Chapter 12 Test Form A

Name ____________

Date ____________

1. Write the first five terms of this recursive sequence. **1.** ____________

$a_1 = 2$ and $a_n = a_{n-1} + n$ for $n > 1$

2. Write the series represented by the summation notation. Then evaluate the sum. **2.** ____________

$$\sum_{i=1}^{4} (i)^2$$

3. Find a formula for the nth term, a_n, of the arithmetic sequence. $a_1 = 5$ and the common difference is 7. **3.** ____________

4. ***Cost of Well Drilling*** A well drilling company charges \$12 for drilling the first foot of a well, \$12.40 for the second foot, \$12.80 for the third foot, and so on. How much would it cost to have them drill a 60-foot well? **4.** ____________

5. Find the common ratio of the geometric sequence. **5.** ____________

$5, -\frac{10}{3}, \frac{20}{9}, -\frac{40}{27}, \ldots$

6. Write a formula for the nth term, a_n, of the geometric sequence. **6.** ____________

$2, 3, \frac{9}{2}, \frac{27}{4}, \ldots$

7. For a geometric sequence the second term is 4 and the fourth term is $\frac{4}{9}$. Write the formula for a_n. Assume $r > 0$. **7.** ____________

8. Evaluate the sum of the finite geometric series. Round the result to two decimal places. **8.** ____________

$$\sum_{n=1}^{10} 4\left(\frac{3}{4}\right)^{n-1}$$

Mid-Chapter 12 Test Form B

Name ______________________

Date ______________________

1. Write the first five terms of this recursive sequence.

$a_1 = 3$ and $a_n = a_{n-1} - n$ for $n > 1$

1. ______________________

2. Write the series represented by the summation notation. Then evaluate the sum.

$$\sum_{i=1}^{4}(3i - 1)$$

2. ______________________

3. Find a formula for the nth term, a_n, of the arithmetic sequence. $a_1 = 8$ and the common difference is 9.

3. ______________________

4. ***Cost of Well Drilling*** A well drilling company charges \$10 for drilling the first foot of a well, \$10.50 for the second foot, \$11.00 for the third foot, and so on. How much would it cost to have them drill a 50-foot well?

4. ______________________

5. Find the common ratio of the geometric sequence.

$16, -12, 9, -\frac{27}{4}, \ldots$

5. ______________________

6. Write a formula for the nth term, a_n, of the geometric sequence.

$2, 5, \frac{25}{2}, \frac{125}{4}, \ldots$

6. ______________________

7. For a geometric sequence the second term is 3 and the fourth term is 12. Write the formula for a_n. Assume $r > 0$.

7. ______________________

8. Evaluate the sum of the finite geometric series. Round the result to two decimal places.

$$\sum_{n=1}^{10} 5\left(\frac{2}{3}\right)^{n-1}$$

8. ______________________

Chapter 12 Test Form A

(Page 1 of 3 pages)

Name ____________________

Date ____________________

1. Write the first five terms of the sequence. Begin with $n = 1$. 1. ____________

$$a_n = \frac{2^n}{n!}$$

2. Write the first five terms of the recursive sequence. 2. ____________

$$a_1 = 5;\ a_n = a_{n-1} + 2$$

3. Write the series represented by the summation notation. Then evaluate the sum. 3. ____________

$$\sum_{k=1}^{4}(k^2 - 1)$$

4. Find the common difference of the arithmetic sequence. 4. ____________

$\frac{1}{2}, \frac{2}{3}, \frac{5}{6}, 1, \ldots$

5. Find a formula for the nth term, a_n, of the arithmetic sequence. $a_1 = -7$ and the common difference is $\frac{5}{2}$. 5. ____________

6. For an arithmetic sequence the common difference is 5 and the sixth term is 40. What is the first term? 6. ____________

7. Find the sum of the first 25 terms of the arithmetic sequence. 7. ____________

8, 15, 22, 29, . . .

 Name ____________

8. Evaluate the sum. **8.** ____________

$$\sum_{i=1}^{30}\left(\tfrac{2}{3}i+4\right)$$

9. Identify the sequence as arithmetic, geometric, or neither. **9.** ____________

1, 1, 2, 6, 24, 120, ...

10. Find the common ratio of the geometric sequence. **10.** ____________

2, −8, 32, −128, ...

11. $a_1 = 5$ and $r = -\frac{1}{2}$ for a geometric sequence. Find a_5. **11.** ____________

12. Write a formula for the nth term of the geometric sequence. **12.** ____________

$24,\ -18,\ \frac{27}{2},\ -\frac{81}{8},\ldots$

13. Find the sum of the finite geometric series. Round your result to two decimal places. **13.** ____________

$$\sum_{n=1}^{10}4\left(\tfrac{2}{3}\right)^{n-1}$$

14. Which of the infinite geometric series does not have a sum? **14.** ____________

a. $\sum_{n=0}^{\infty}5\left(\frac{3}{4}\right)^{n}$ **b.** $\sum_{n=1}^{\infty}5\left(\frac{4}{3}\right)^{n-1}$

15. Find the sum, if it exists. 15. ____________

$$\sum_{n=1}^{\infty} 7\left(\tfrac{2}{5}\right)^{n-1}$$

16. Find the sum, if it exists. 16. ____________

$$\sum_{n=1}^{\infty} (0.01)^n$$

17. Evaluate the binomial coefficient. $\binom{8}{4}$ 17. ____________

18. Expand $(2s - 3t)^3$. 18. ____________

19. What is the coefficient of the x^7y^3 term in the expansion of $(x + y)^{10}$? 19. ____________

20. ***Balance of an Annuity*** The balance, A (in dollars), of an increasing annuity with equal periodic deposits of D dollars made n times per year is given by 20. ____________

$$A = D\left[\frac{(1+i)^{nt} - 1}{i}\right]$$

where $i = \dfrac{r}{n}$, r is the annual interest rate, and t is the elapsed time in years. Find the value of such an annuity at 6% annual rate, where \$100 per month has been deposited for 10 years.

Chapter 12 Test

Form B

Name ______________________

Date ______________________

1. Write the first five terms of the sequence. Begin with $n = 1$.

$$a_n = \frac{n^2}{2^n}$$

1. ______________

2. Write the first five terms of the recursive sequence.

$a_1 = 6;\ a_n = a_{n-1} - 3$

2. ______________

3. Write the series represented by the summation notation. Then evaluate the sum.

$$\sum_{k=1}^{4}(k^2 + 2)$$

3. ______________

4. Find the common difference of the arithmetic sequence.

$\frac{1}{2}, \frac{1}{6}, -\frac{1}{6}, -\frac{1}{2}, \ldots$

4. ______________

5. Find a formula for the nth term, a_n, of the arithmetic sequence. $a_1 = 8$ and the common difference is $\frac{2}{3}$.

5. ______________

6. For an arithmetic sequence the common difference is 6 and the fifth term is 35. What is the first term?

6. ______________

7. Find the sum of the first 30 terms of the arithmetic sequence.

9, 17, 25, 33, . . .

7. ______________

Name ____________

8. Evaluate the sum. **8.** ____________

$$\sum_{i=1}^{20} \left(\tfrac{3}{2}i + 4\right)$$

9. Identify the sequence as arithmetic, geometric, or neither. **9.** ____________

1, 1, 2, 3, 5, 8, 13, . . .

10. Find the common ratio of the geometric sequence. **10.** ____________

$-\frac{1}{3}$, 1, −3, 9, . . .

11. $a_1 = 4$ and $r = -\frac{1}{2}$ for a geometric sequence. Find a_6. **11.** ____________

12. Write a formula for the nth term of the geometric sequence. **12.** ____________

48, −32, $\frac{64}{3}$, $-\frac{128}{9}$, . . .

13. Find the sum of the finite geometric series. Round your result to two decimal places. **13.** ____________

$$\sum_{n=1}^{10} 4\left(\tfrac{3}{2}\right)^{n-1}$$

14. Which of the infinite geometric series does not have a sum? **14.** ____________

a. $\sum_{n=0}^{\infty} 6\left(\frac{4}{5}\right)^n$ **b.** $\sum_{n=1}^{\infty} 6\left(\frac{5}{4}\right)^{n-1}$

 Name ______________________

15. Find the sum, if it exists. **15.** ______________

$$\sum_{n=1}^{\infty} 12\left(\tfrac{1}{3}\right)^{n-1}$$

16. Find the sum, if it exists. **16.** ______________

$$\sum_{n=1}^{\infty} (0.02)^n$$

17. Evaluate the binomial coefficient. $\binom{9}{5}$ **17.** ______________

18. Expand $(3s - 2t)^3$. **18.** ______________

19. What is the coefficient of the x^6y^4 term in the expansion of $(x + y)^{10}$? **19.** ______________

20. ***Balance in an Annuity*** The balance, A (in dollars), of an increasing annuity with equal periodic deposits of D dollars made n times per year is given by **20.** ______________

$$A = D\left[\frac{(1 + i)^{nt} - 1}{i}\right]$$

where $i = \frac{r}{n}$, r is the annual interest rate, and t is the elapsed time in years. Find the value of such an annuity at 7% annual rate, where $100 per month has been deposited for 10 years.

Chapter 12 Test Form C

Name ____________________

Date ____________________

1. Write the first five terms of the sequence. Begin with $n = 1$. 1. ____________

   $a_n = \frac{n^3}{n!}$

2. Write the first five terms of the recursive sequence. 2. ____________

   $a_1 = 1;\ a_n = a_{n-1} + n^2$

3. Write the series represented by the summation notation. Then evaluate the sum. 3. ____________

   $\sum_{n=1}^{4} (n! - 1)$

4. Find the common difference of the arithmetic sequence. 4. ____________

   $-\frac{3}{2},\ 1,\ \frac{7}{2},\ 6, \ldots$

5. Find a formula for the nth term, a_n, of the arithmetic sequence. $a_1 = \frac{1}{2}$ and the common difference is $\frac{1}{6}$. 5. ____________

6. For an arithmetic sequence the common difference is -7 and the seventh term is 40. What is the first term? 6. ____________

7. Find the sum of the first 30 terms of the arithmetic sequence. 7. ____________

   5, 14, 23, 32, ...

 Name ______________________

8. Evaluate the sum. **8.** ______________

$$\sum_{i=1}^{30}\left(\tfrac{1}{4}i+3\right)$$

9. Identify the sequence as arithmetic, geometric, or neither. **9.** ______________

1, 4, 9, 16, 25, . . .

10. Find the common ratio of the geometric sequence. **10.** ______________

125, 75, 45, 27, $\frac{81}{5}$, . . .

11. For a geometric sequence, $a_2 = -4$ and $a_5 = \frac{4}{27}$. Find a_1 and a_6. **11.** ______________

12. Write a formula for the nth term of the geometric sequence. **12.** ______________

3, -2, $\frac{4}{3}$, $-\frac{8}{9}$, . . .

13. Find the sum of the finite geometric series. Round your result to two decimal places. **13.** ______________

$$\sum_{n=1}^{10} 5\left(\tfrac{3}{4}\right)^{n-1}$$

14. Which of the infinite geometric series does not have a sum? **14.** ______________

a. $\sum_{n=0}^{\infty}(0.99)^n$ **b.** $\sum_{n=1}^{\infty}(1.01)^{n-1}$

15. Find the sum, if it exists. 15. ____________

$$\sum_{n=1}^{\infty} 4\left(\tfrac{3}{7}\right)^{n-1}$$

16. Find the sum, if it exists. 16. ____________

$$\sum_{n=1}^{\infty} (0.99)^n$$

17. Evaluate the binomial coefficient. $\binom{12}{5}$ 17. ____________

18. Expand $(A - 2B)^4$. 18. ____________

19. What is the coefficient of the x^5y^5 term in the expansion of $(x + y)^{10}$? 19. ____________

20. ***Balance in an Annuity*** In an increasing annuity, the periodic deposit, D (in dollars), made n times per year to produce a balance, A (in dollars), after t years is given by 20. ____________

$$D = A\left[\frac{i}{(1+i)^{nt} - 1}\right]$$

where $i = \frac{r}{n}$ and r is the annual interest rate. Find the monthly deposit which must be made for such an annuity if the annual interest rate is 8% and the balance is expected to be \$10,000 after 5 years.

Cumulative Test

Chapters 7–12

(Page 1 of 8 pages)

Name ______________________

Date ______________________

1. Simplify. $\dfrac{4^{-2}x^3y^{-3}}{2^3(xy^{-2})^2} \cdot \dfrac{(x^3y^3)^{-1}}{(8x)^{-3}}$ **1.** ______________

2. Solve for x. $3^{-2} \cdot 9^x = 27^2$ **2.** ______________

3. Evaluate. $\left(\frac{16}{81}\right)^{-3/4}$ **3.** ______________

4. ***Balance in an Account*** $2500 is deposited in an account paying 7.5% annual interest, compounded quarterly. Find its value after 5 years. **4.** ______________

5. Use a calculator to evaluate $15^{5/4}$ to three decimal places. **5.** ______________

6. Evaluate $\sqrt[6]{15181}$ to three decimal places. **6.** ______________

7. ***Depreciated Value*** New farm equipment costs $60,000. If it depreciates 11% each year, find its value to the nearest $100 after ten years. **7.** ______________

Name ______________________

8. Simplify. $\sqrt{32} - \sqrt{98} + 3\sqrt{50}$ **8.** ______________

9. Expand the expression. $\ln \sqrt[4]{\dfrac{x^2 z}{y^3}}$ **9.** ______________

10. Use a calculator to evaluate $e^{-0.417}$ to three decimal places. **10.** ______________

11. Solve for x. $\log_3(2x + 3) - \log_3 4 = \log_3 x$ **11.** ______________

12. Sketch the graph of the function.

$f(x) = 2(e^{-x} - 1)$

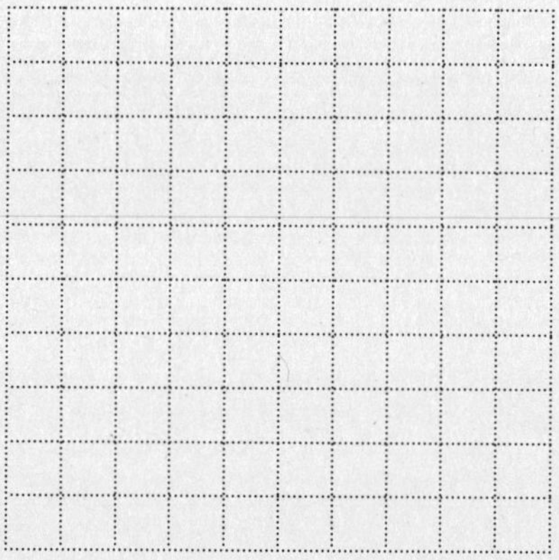

12. *Use graph at left.*

13. Evaluate $\log_7 92$ to three decimal places. **13.** ______________

14. Sketch the graph of the function.

$f(x) = \ln(x + 1)$

14. *Use graph at left.*

Name ______________________

15. Perform the indicated operations. **15.** ______________

$(5x + 4)(x^2 - 1) - 3(2 - x^2)$

16. Factor completely with respect to the integers. **16.** ______________

$x^3 + 3x^2 - 9x - 27$

17. Sketch the graph of the function. **17.** *Use graph at left.*

$g(x) = (x - 2)^3$

18. Perform the division. Write the answer in fractional form. **18.** ______________

$(2x^3 + 10x^2 - 11x + 9) \div (x + 6)$

19. Write a polynomial whose graph has the given x-intercepts and has a leading coefficient of 1. **19.** ______________

$(-1,\ 0), (2,\ 0), (-3,\ 0)$

 Name ____________

20. Find all real zeros of the function.

$$g(x) = x^3 + 2x^2 - 5x - 10$$

20. ____________

21. Find the *range*, the *mean*, and the *standard deviation* of the set of data. Round your result to two decimal places.

52.1, 47.8, 49.4, 53.7, 45.1, 49.2, 47.3, 48.5, 48.3, 46.3

21. ____________

22. Sketch the graph of the function. Identify any asymptotes.

$$f(x) = \frac{1 - x}{x + 1}$$

22. ____________ *Use graph at left.*

23. x and y vary inversely. $x = 12$ when $y = \frac{4}{3}$. Find an equation that relates the variables.

23. ____________

 Name ____________

24. Perform the indicated operations and simplify. **24.** ____________

$$\frac{x^2+3x-10}{x^3-4x} \div \frac{x^2+5x}{x^2+2x}$$

25. Solve the equation. **25.** ____________

$$\frac{x}{45} - \frac{2}{x} = \frac{1}{5}$$

26. ***Average Cost*** Startup costs for producing a product is \$12,000. Thereafter, each item costs \$6 to produce. How many must be produced to bring the *average cost* per item down to \$26? **26.** ____________

27. Simplify the complex fraction. **27.** ____________

$$\frac{\frac{5}{x+3} - \frac{2}{3}}{\frac{1}{6} + \frac{2}{x+3}}$$

28. ***Total Interest*** Use the formula as given to find the monthly payment required to pay off a \$2000 loan at 12% annual interest in 3 years. What is the total interest cost of the loan? **28.** ____________

$$M = P\left[\frac{i}{1-\left(\frac{1}{1+i}\right)^{12t}}\right]$$

($i = \frac{r}{12}$; r is the annual rate; t is in years)

 Name ____________

29. A parabola has vertex at (0, 0) and focus at (0, 2). Write its equation and sketch the graph.

29. ____________
Use graph at left.

30. A circle has center at (2, −4) and passes through the origin. Write its equation.

30. ____________

31. Find the *vertices*, the *foci*, and the *eccentricity* of the ellipse. Sketch the graph.

$$\frac{x^2}{36} + \frac{y^2}{16} = 1$$

31. ____________
Use graph at left.

32. A hyperbola has vertices at $(0, \pm 3)$ and foci at $(0, \pm 5)$. Write the equation of the hyperbola, list the asymptotes, and sketch the graph.

32. ____________
Use graph at left.

 Name ____________________

33. Classify the conic as a circle, an ellipse, a hyperbola, or a parabola.
$$7x^2 + 12y^2 - 14x + 24y - 28 = 0$$

33. ____________________

34. Write the first five terms of the sequence. Begin with $n = 1$.
$$a_n = \frac{n^2 + 1}{2^n}$$

34. ____________________

35. Find a formula for the nth term, a_n, of the arithmetic sequence.
$a_1 = \frac{5}{4}$ and the common difference is $\frac{1}{2}$.

35. ____________________

36. Evaluate the sum $\sum_{n=1}^{40} \left(\frac{4}{3}n + \frac{2}{3}\right)$.

36. ____________________

37. $a_1 = 96$ and $r = -\frac{1}{2}$ for a geometric sequence.
Find a formula for a_n and evaluate a_{12}.

37. ____________________

38. Find the sum of the finite geometric series.

$$\sum_{n=1}^{6} 9\left(\tfrac{5}{6}\right)^{n-1}$$

38. ____________

39. Find the sum, if it exists.

$$\sum_{n=1}^{\infty} 100\left(\tfrac{1}{5}\right)^{n-1}$$

39. ____________

40. Evaluate the binomial coefficient. $\binom{10}{5}$

40. ____________

41. Expand $(2x - 3)^4$.

41. ____________

Mid-Chapter 13 Test Form A

Name ______________________

Date ______________________

1. Find $\sin\theta$ when $\cos\theta = \frac{1}{3}$. Do not use a calculator and assume θ is an acute angle. 1. ____________

2. Find $\sec\theta$ when $\tan\theta = 2$. Do not use a calculator and assume θ is an acute angle. 2. ____________

3. Use a calculator to evaluate $\csc 23°$ to three decimal places. 3. ____________

4. Use a calculator to evaluate $\cot 72°$ to three decimal places. 4. ____________

5. Which of the angles are *coterminal*? 5. ____________

   $40°,\ 400°,\ -400°$

6. Find the *complement* of $\frac{\pi}{6}$ radians. 6. ____________

7. Find the *supplement* of $123°$. 7. ____________

8. Convert $\frac{3\pi}{5}$ radians to degree measure. 8. ____________

9. The point $(1,\ -\sqrt{3})$ is on the terminal side of an angle θ. Find $\cos\theta$. 9. ____________

10. Decide whether the statement is correct or not correct. 10. ____________

    $$\sin\frac{\pi}{4} = \sin\frac{3\pi}{4}$$

Mid-Chapter 13 Test Form B

Name ______________________

Date ______________________

1. Find $\cos\theta$ when $\sin\theta = \frac{2}{3}$. Do not use a calculator and assume θ is an acute angle. 1. ______________

2. Find $\tan\theta$ when $\sec\theta = \sqrt{7}$. Do not use a calculator and assume θ is an acute angle. 2. ______________

3. Use a calculator to evaluate $\csc 68°$ to three decimal places. 3. ______________

4. Use a calculator to evaluate $\cot 29°$ to three decimal places. 4. ______________

5. Which of the angles are *coterminal*? 5. ______________

   $-320°,\ 40°,\ -40°$

6. Find the *complement* of $37°$. 6. ______________

7. Find the *supplement* of $\frac{3\pi}{4}$ radians. 7. ______________

8. Convert $\frac{3\pi}{8}$ radians to degree measure. 8. ______________

9. The point $(-5,\ 12)$ is on the terminal side of an angle θ. Find $\tan\theta$. 9. ______________

10. Decide whether the statement is correct or not correct. 10. ______________

    $\cos\frac{\pi}{4} = \cos\frac{3\pi}{4}$

Chapter 13 Test Form A

(Page 1 of 3 pages)

Name ______________________

Date ______________________

In 1–5, evaluate without using a calculator.

1. csc 45° **1.** ______________

2. sin 30° **2.** ______________

3. cot 60° **3.** ______________

4. Find $\sin\theta$ when θ is an acute angle and $\cos\theta = \frac{1}{5}$. **4.** ______________

5. Find $\sec\theta$ when θ is an acute angle and $\cot\theta = \frac{1}{2}$. **5.** ______________

In 6–10, use a calculator. Round to three decimal places.

6. cos 12° **6.** ______________

7. tan 69° **7.** ______________

8. sin 12° **8.** ______________

9. sec 27° **9.** ______________

10. csc 20° **10.** ______________

 Name ______________

11. Which of the angles are coterminal? 11. ______________

$10°$, $80°$, $370°$

12. Find the complement of $74°$. 12. ______________

13. Find the complement of $\frac{\pi}{6}$ radians. 13. ______________

14. Find the supplement of $126°$. 14. ______________

15. Find the supplement of $\frac{\pi}{12}$ radians. 15. ______________

16. Convert $48°$ to radian measure. 16. ______________

17. Convert $\frac{7\pi}{15}$ to degree measure. 17. ______________

18. The point $(-3, 3)$ is on the terminal side of an angle θ. Find $\cos\theta$. 18. ______________

19. Evaluate. $\arccos\left(-\frac{\sqrt{2}}{2}\right)$

19. ____________

20. ***Geometry*** Solve for x in the triangle shown.

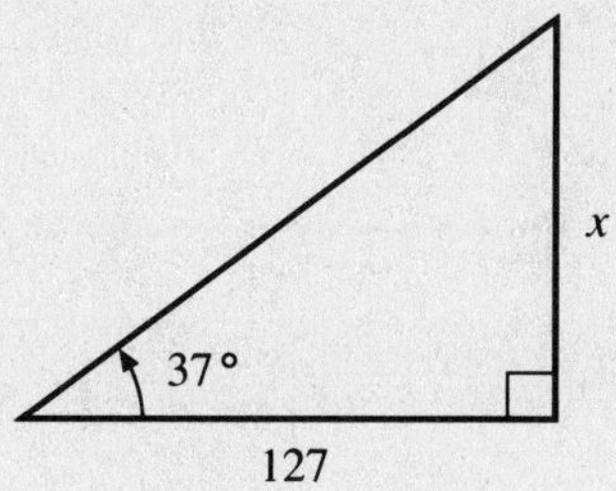

20. ____________

21. ***Geometry*** Solve for y in the triangle shown.

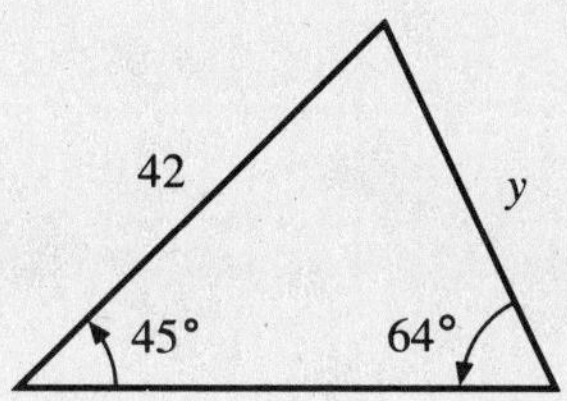

21. ____________

22. ***Geometry*** Find the area of the triangle and the length of the remaining side.

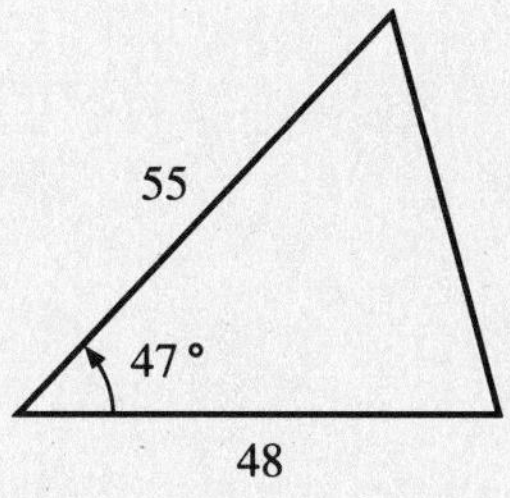

22. ____________

Chapter 13 Test **Form B**

Name ______________________

Date ______________________

In 1–5, evaluate without using a calculator.

1. $\cot 45^\circ$ **1.** ______________

2. $\cos 60^\circ$ **2.** ______________

3. $\sec 30^\circ$ **3.** ______________

4. Find $\cos\theta$ when θ is an acute angle and $\sin\theta = \frac{1}{3}$. **4.** ______________

5. Find $\csc\theta$ when θ is an acute angle and $\tan\theta = \frac{2}{3}$. **5.** ______________

In 6–10, use a calculator. Round to three decimal places.

6. $\tan 17^\circ$ **6.** ______________

7. $\cos 69^\circ$ **7.** ______________

8. $\sec 12^\circ$ **8.** ______________

9. $\sin 27^\circ$ **9.** ______________

10. $\cot 20^\circ$ **10.** ______________

Name ____________________

11. Which of the angles are coterminal? **11.** ____________

$30°$, $-330°$, $-30°$

12. Find the complement of $26°$. **12.** ____________

13. Find the complement of $\frac{\pi}{4}$ radians. **13.** ____________

14. Find the supplement of $36°$. **14.** ____________

15. Find the supplement of $\frac{3\pi}{8}$ radians. **15.** ____________

16. Convert $36°$ to radian measure. **16.** ____________

17. Convert $\frac{2\pi}{3}$ radians to degree measure. **17.** ____________

18. The point $(-1, \sqrt{3})$ is on the terminal side of an angle θ. Find $\cos\theta$. **18.** ____________

Name ______________________

19. Evaluate. $\arctan\left(-\frac{\sqrt{3}}{3}\right)$

19. ______________________

20. ***Geometry*** Solve for x in the triangle shown.

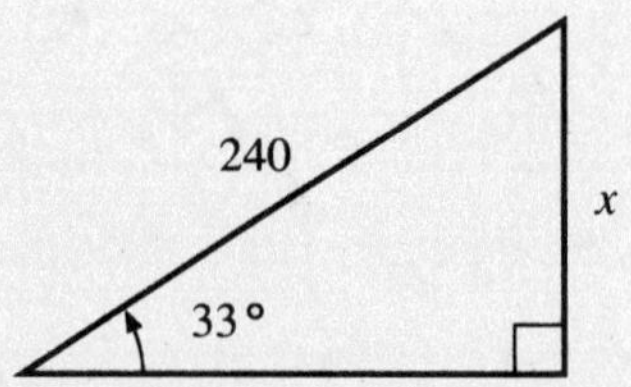

20. ______________________

21. ***Geometry*** Solve for y in the triangle shown.

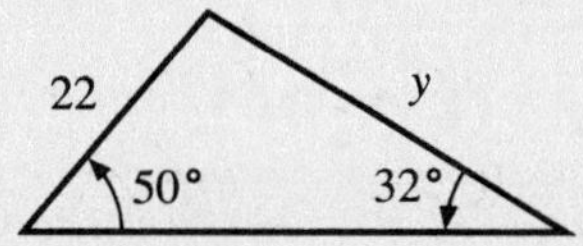

21. ______________________

22. ***Geometry*** Find the area of the triangle and the length of the remaining side.

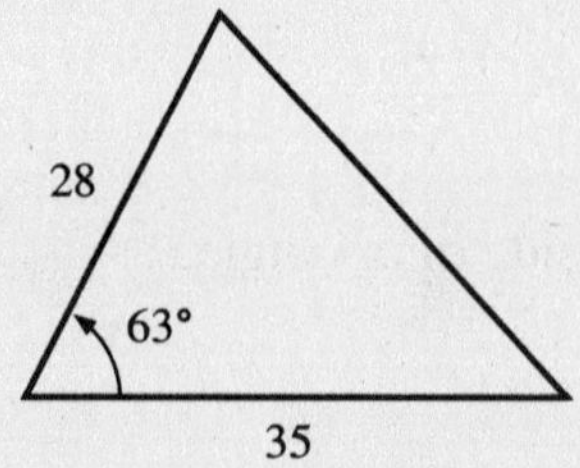

22. ______________________

Chapter 13 Test Form C

(Page 1 of 3 pages)

Name ______________________

Date ______________________

In 1–5, evaluate without using a calculator.

1. $\csc 30°$ 　　**1.** ______________

2. $\cos 45°$ 　　**2.** ______________

3. $\tan 60°$ 　　**3.** ______________

4. Find $\sin\theta$ when θ is an acute angle and $\sec\theta = 4$. 　　**4.** ______________

5. Find $\cos\theta$ when θ is an acute angle and $\cot\theta = \frac{3}{2}$. 　　**5.** ______________

In 6–10, use a calculator. Round to three decimal places.

6. $\csc 8°$ 　　**6.** ______________

7. $\tan 42°$ 　　**7.** ______________

8. $\cos 80°$ 　　**8.** ______________

9. $\sec 36°$ 　　**9.** ______________

10. $\sin 69°$ 　　**10.** ______________

Name ____________________

11. Which of the angles are coterminal? **11.** ____________

$\frac{\pi}{4}, \frac{3\pi}{4}, -\frac{7\pi}{4}$

12. Find the complement of 45°. **12.** ____________

13. Find the complement of $\frac{4\pi}{9}$ radians. **13.** ____________

14. Find the supplement of 4°. **14.** ____________

15. Find the supplement of $\frac{5\pi}{9}$ radians. **15.** ____________

16. Convert 18° to radian measure. **16.** ____________

17. Convert $\frac{5\pi}{12}$ radians to degree measure. **17.** ____________

18. The point $(-1, -2)$ is on the terminal side of an angle θ. Find $\sec\theta$. **18.** ____________

 Name ____________________

19. Evaluate arctan(−0.314). Give the result in radians to three decimal places. **19.** ____________

20. ***Geometry*** Solve for x in the triangle shown. **20.** ____________

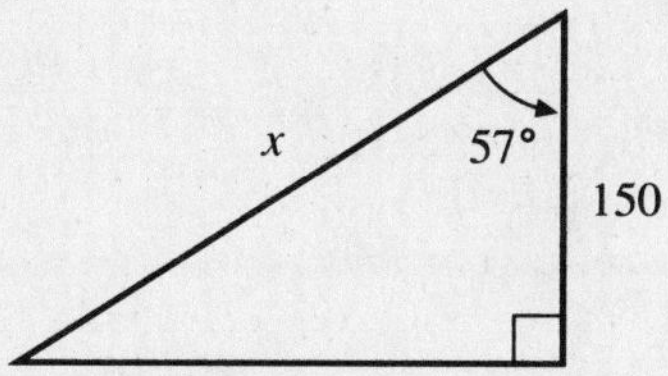

21. ***Geometry*** Solve for y in the triangle shown. **21.** ____________

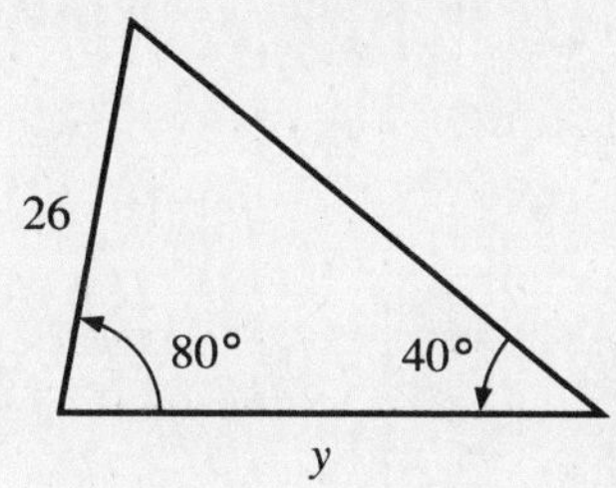

22. ***Geometry*** Find the area of the triangle and the length of the remaining side. **22.** ____________

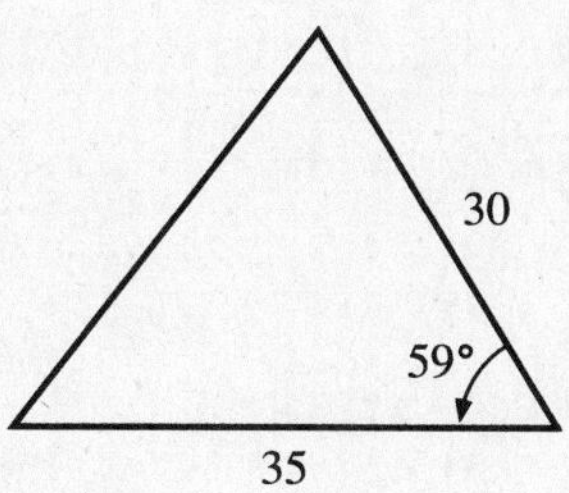

Mid-Chapter 14 Test Form A

Name ______________________

Date ______________________

1. Find the *amplitude* and *period* of the graph. 1. ______________

   $y = \frac{3}{2} \sin 2x$

2. Sketch one cycle of the graph of the function. 2. Use graph at left.

   $f(x) = 4 \cos \frac{\pi x}{2}$

3. Write an equation of the form $y = a \cos bx$ with amplitude 6 and period $\frac{3\pi}{2}$. 3. ______________

4. The graph of $y = 7 \sin 2x$ is drawn and then shifted down two units. Write the equation of the resulting graph. 4. ______________

5. Sketch one cycle of the graph of the function. 5. Use graph at left.

   $y = -1 + 2 \cos(x + \frac{\pi}{2})$

6. Simplify the expression. $(\sin x)(1 + \cot^2 x)$ 6. ______________

7. Verify the identity. 7. ______________

   $\cos^4 x - \sin^4 x = \cos^2 x - \sin^2 x$

Mid-Chapter 14 Test Form B

Name ______________________

Date ______________________

1. Find the *amplitude* and *period* of the graph.

$$y = 5\cos\frac{\pi}{2}x$$

1. ______________________

2. Sketch one cycle of the graph of the function.

$$y = 2\sin 2x$$

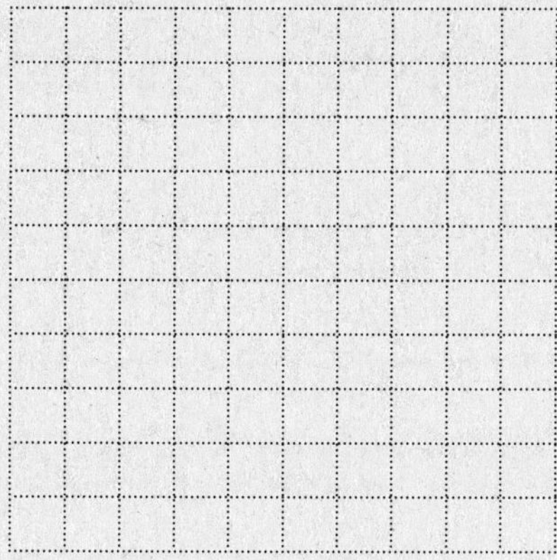

2. *Use graph at left.*

3. Write an equation of the form $y = a\sin bx$ with amplitude 3 and period 3.

3. ______________________

4. The graph of $y = -2\cos 3x$ is drawn and then shifted two units to the left. Write the equation of the resulting graph.

4. ______________________

5. Sketch one cycle of the graph of the function.

$$y = 1 - 2\sin(x - \frac{\pi}{2})$$

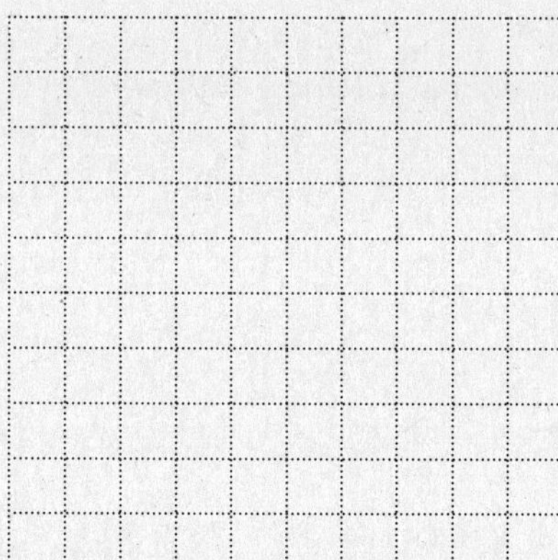

5. *Use graph at left.*

6. Simplify the expression. $\dfrac{\cos x}{\cot(-x)}$

6. ______________________

7. Verify the identity.

$$\tan^2 x + \sec^2 x = 2\sec^2 x - 1$$

7. ______________________

Chapter 14 Test Form A

(Page 1 of 3 pages)

Name ________________

Date ________________

1. Find the *amplitude* and *period* of the graph. **1.** ____________

$y = 3\sin 2\pi x$

2. Sketch one cycle of the graph of the function. **2.** *Use graph at left.*

$f(x) = 4\cos\dfrac{\pi x}{3}$

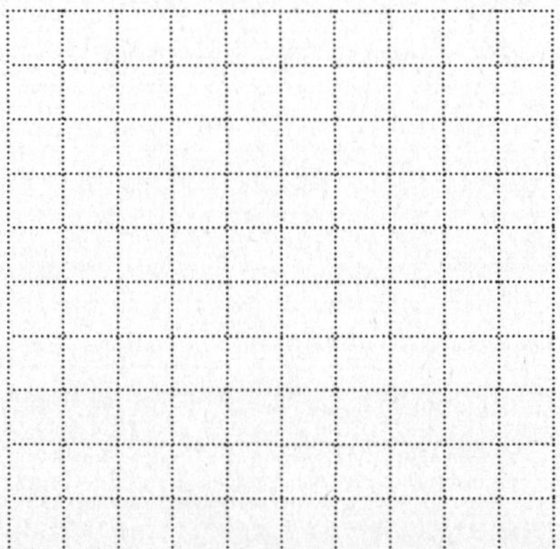

3. Write an equation of the form $y = a\cos bx$ with amplitude 0.6 and period $\dfrac{\pi}{2}$. **3.** ____________

4. Write two x-values at which the function has a *maximum*. **4.** ____________

$f(x) = 4\cos 2x$

5. Which graph oscillates more frequently? **5.** ____________

a. $y = \frac{1}{2}\sin 4x$ **b.** $y = \frac{1}{4}\sin 2x$

6. The graph of $y = 6\sin\dfrac{\pi x}{3}$ is drawn and then shifted up two units. Write the equation of the resulting graph. **6.** ____________

7. Sketch one cycle of the graph of the function.

$y = \frac{1}{2} - \frac{5}{2}\cos\frac{\pi x}{2}$

7. *Use graph at left.*

8. Simplify the expression.

$\sin\left(\frac{\pi}{2} - x\right)\sec x$

8. ______________________

9. Simplify the expression.

$\csc x \cos(-x)$

9. ______________________

10. Verify the identity.

$\frac{\sin x}{1 + \cos x} = \frac{1 - \cos x}{\sin x}$

10. ______________________

11. Solve the equation for $0 \leq x < 2\pi$.

$4\sin^2 x - 1 = 0$

11. ______________________

12. Simplify the expression. Do not evaluate it.

$\sin 30^\circ \cos 40^\circ + \cos 30^\circ \sin 40^\circ$

12. ______________________

13. Find the intersection points of the two graphs in the interval $0 \le x \le 2\pi$. 13. ____________

$$y = 2\sin^2 x;\; y = \sin x$$

14. Find the *exact* value of the expression 14. ____________

$$\cos 75°.$$

15. Solve the equation for $0 \le x < 2\pi$. 15. ____________

$$\sin 2x - \cos x = 0$$

16. Find the *exact* value of the expression. 16. ____________

$$\cos\left(\frac{\pi}{12}\right)$$

Chapter 14 Test Form B

(Page 1 of 3 pages)

Name ______________________

Date ______________________

1. Find the *amplitude* and *period* of the graph.

$y = -3 \cos \pi x$

1. ______________________

2. Sketch one cycle of the graph of the function.

$f(x) = 6 \sin \frac{x}{2}$

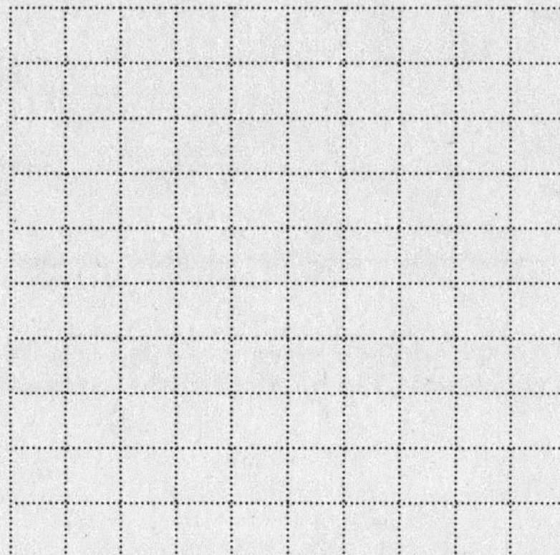

2. *Use graph at left.*

3. Write an equation of the form $y = a \sin bx$ with amplitude $\frac{2}{3}$ and period 12.

3. ______________________

4. Write two x-values at which the function has a *minimum*.

$f(x) = 4 \sin 2x$

4. ______________________

5. Which graph oscillates more frequently?

a. $y = \frac{1}{4} \cos 2x$ **b.** $y = \frac{1}{2} \cos 4x$

5. ______________________

6. The graph of $y = -8 \cos 6\pi x$ is drawn and then shifted down three units. Write the equation of the resulting graph.

6. ______________________

 Name ____________________

7. Sketch one cycle of the graph of the function.

$y = 1 - \frac{3}{2} \sin \frac{x}{2}$

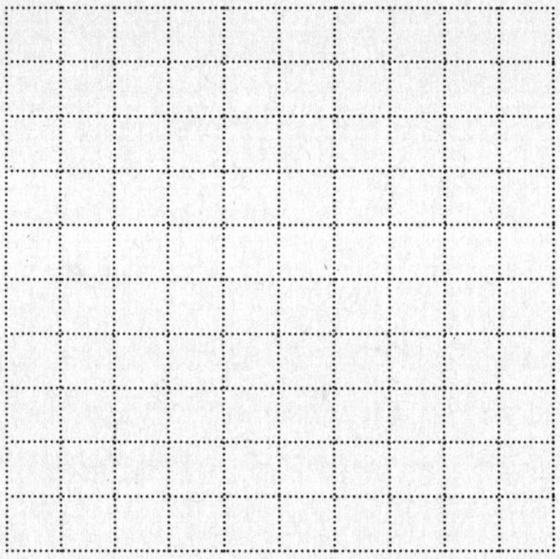

7. *Use graph at left.*

8. Simplify the expression.

$\cos\left(\frac{\pi}{2} - x\right) \sec x$

8. ____________________

9. Simplify the expression.

$\frac{\sin^2(-x)}{\tan^2 x}$

9. ____________________

10. Verify the identity.

$\tan^2 x(\csc^2 x - 1) = 1$

10. ____________________

11. Solve the equation for $0 \le x < 2\pi$.

$4\cos^2 x - 1 = 0$

11. ____________________

12. Simplify the expression. Do not evaluate it.

$\cos 100° \cos 20° + \sin 100° \sin 20°$

12. ____________________

Name ______________________

13. Find the intersection points of the two graphs in the interval $0 \le x \le 2\pi$.

$y = \cos x;\ \ y = 2\cos^2 x$

13. ______________________

14. Find the *exact* value of the expression

$\cos 15°$.

14. ______________________

15. Solve the equation for $0 \le x < 2\pi$.

$\cos 2x + \cos x = 0$

15. ______________________

16. Find the *exact* value of the expression.

$\sin\left(\frac{\pi}{8}\right)$

16. ______________________

Chapter 14 Test Form C

Name ______________________

Date ______________________

1. Find the *amplitude* and *period* of the graph. **1.** ______________

$y = -2\cos 6x$

2. Sketch one cycle of the graph of the function. **2.** *Use graph at left.*

$f(x) = 3\sin\dfrac{x}{2}$

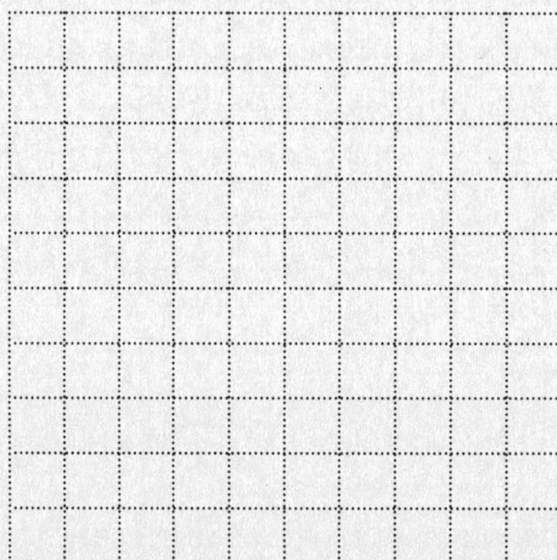

3. Write an equation of the form $y = a\cos bx$ with amplitude $\frac{1}{6}$ and period $\dfrac{\pi}{3}$. **3.** ______________

4. Write two x-values at which the function has a *minimum*. **4.** ______________

$f(x) = 4\cos 4x$

5. Which graph oscillates more frequently? **5.** ______________

a. $y = 2\sin\dfrac{\pi x}{4}$ **b.** $y = 4\sin\dfrac{\pi x}{2}$

6. The graph of $y = -\dfrac{1}{5}\cos\dfrac{3x}{8}$ is drawn and then shifted two units to the left. Write the equation of the resulting graph. **6.** ______________

 Name ______________________

7. Sketch one cycle of the graph of the function.

$y = -1 + 2\cos \pi x$

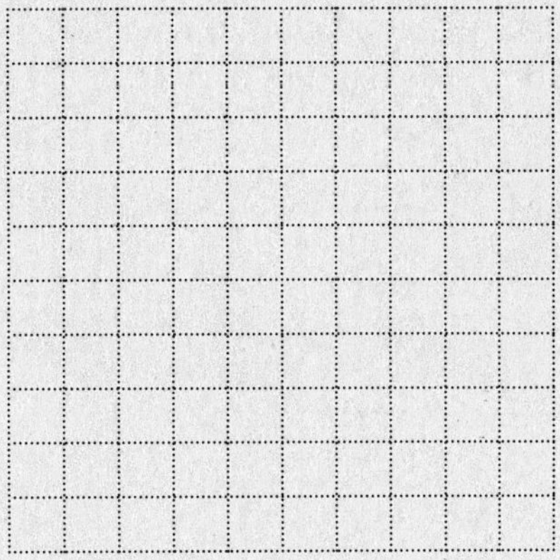

7. *Use graph at left.*

8. Simplify the expression.

$\csc^2\left(\frac{\pi}{2} - x\right) - 1$

8. ______________________

9. Simplify the expression.

$\frac{\cos(-x)}{\sec x} + \sin^2(-x)$

9. ______________________

10. Verify the identity.

$\cos^2 x - \cos^4 x = \sin^2 x - \sin^4 x$

10. ______________________

11. Solve the equation for $0 \le x < 2\pi$.

$2\sin^2 x - \sin x - 1 = 0$

11. ______________________

12. Simplify the expression. Do not evaluate it.

$\cos\frac{\pi}{4}\cos\frac{\pi}{3} - \sin\frac{\pi}{4}\sin\frac{\pi}{3}$

12. ______________________

13. Find the intersection points of the two graphs in the interval $0 \le x \le 2\pi$.

$y = 4\cos^2 x;\ \ y = 4\cos x - 1$

13. ____________

14. Find the *exact* value of the expression

$\tan 15°$.

14. ____________

15. Solve the equation for $0 \le x < 2\pi$.

$\sin 2x \cos x = \sin x$

15. ____________

16. Find the *exact* value of the expression.

$\sin\left(\frac{7\pi}{8}\right)$

16. ____________

Mid-Chapter 15 Test Form A

Name ______________________

Date ______________________

1. ***Balls in a Jar*** A jar contains seven balls numbered 1 to 7. Three are withdrawn at random. How many different combinations are possible? 1. ______________

2. ***Choosing a Committee*** You are a member of a group of 15 people. A three-member committee is chosen at random from the group. How many different committees are possible? 2. ______________

3. ***Forming a Code*** A three-digit code is formed. The first digit is any letter of the alphabet except I and O. The second digit is any integer from 1 through 9. The third digit is any integer from 0 through 9. How many codes are possible? 3. ______________

4. A prime number is divisible only by itself and 1. 1 is not considered a prime. If 3 numbers from 1 through 20 are chosen at random, find the probability that they are all prime numbers. 4. ______________

5. ***Geometry*** A circle is inscribed in a square as shown. A point inside the square is selected at random. Find the probability that the point is also inside the circle. 5. ______________

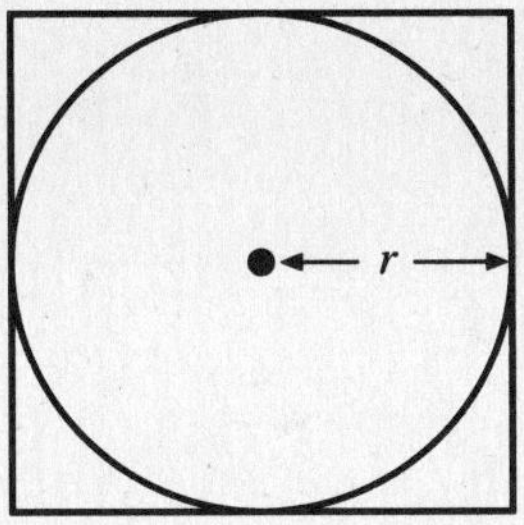

6. ***Drawing Three Cards*** Three cards are drawn from a standard deck of 52 playing cards. Find the probability that they are all red cards (hearts or diamonds). 6. ______________

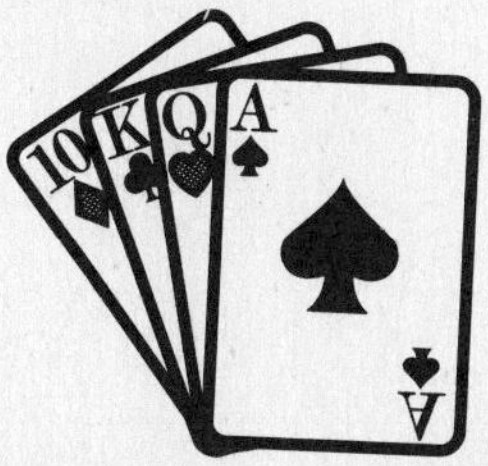

Mid-Chapter 15 Test Form B

Name ______________________

Date ______________________

1. ***Balls in a Jar*** A jar contains eight balls numbered 1 to 8. Three are withdrawn at random. How many different combinations are possible? 1. ______________

2. ***Choosing a Committee*** You are a member of a group of 11 people. A three-member committee is chosen at random from the group. How many different committees are possible? 2. ______________

3. ***Forming a Code*** A three-digit code is formed. The first digit is any letter of the alphabet except vowels (A, E, I, O, U). The second digit is any integer from 0 through 9. The third digit is any of the vowels. How many codes are possible? 3. ______________

4. A prime number is divisible only by itself and 1. 1 is not considered a prime. If 3 numbers from 1 through 20 are chosen at random, find the probability that none are prime numbers. 4. ______________

5. ***Geometry*** A circle is inscribed in a square as shown. A point inside the square is selected at random. Find the probability that the point is not inside the circle. 5. ______________

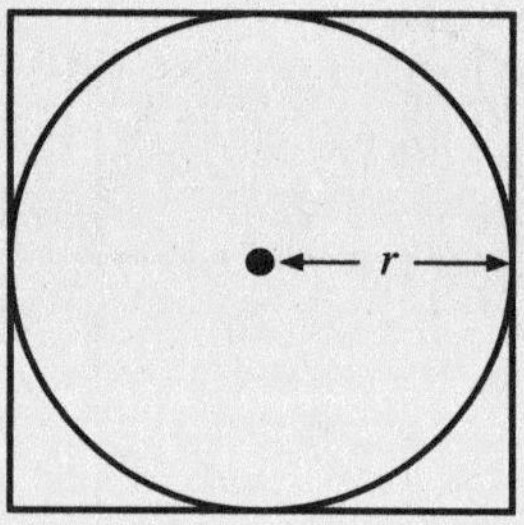

6. ***Drawing Three Cards*** Three cards are drawn from a standard deck of 52 playing cards. Find the probability that they are all face cards (**Q**, **K**, or **J**). 6. ______________

Chapter 15 Test	**Form A**

(Page 1 of 3 pages) **Name** ______________________

1. ***Choosing a Committee*** A four-person committee is chosen at random from a group of 15 people. How many different committees are possible? 1. ________

2. ***Permuting Letters*** How many ways can the letters in the word LUCKY be permuted? 2. ________

3. ***Drawing Cards*** Three cards are drawn from a standard deck of playing cards. How many different such combinations are possible containing exactly one ace? 3. ________

4. How many different three-digit numbers are possible if the first two digits are the same but the third is different? 4. ________

5. ***Drawing One Card*** A card is drawn from a standard deck of playing cards. Find the probability that it is not a face card (**J**, **Q**, or **K**). 5. ________

6. ***Tossing a Coin*** A coin is tossed 5 times and heads (H) or tails (T) is recorded as the result in each case. How many different outcomes are possible for the sequence of 5 tosses? 6. ________

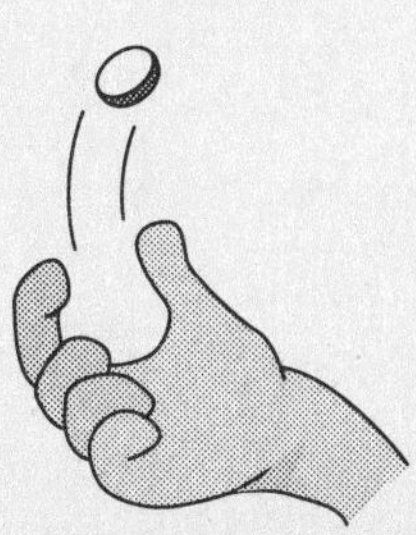

 Name ____________________

7. ***Assigning Seats*** You, your friend, and four others are assigned seats at random in a row of 6 chairs. How many ways can this be done and what is the probability that you and your friend will be seated next to each other? **7.** ________

8. ***Geometry*** Half of a circle is inside a square and half is outside, as shown. If a point is selected at random inside the square, find the probability the point is also inside the circle. **8.** ________

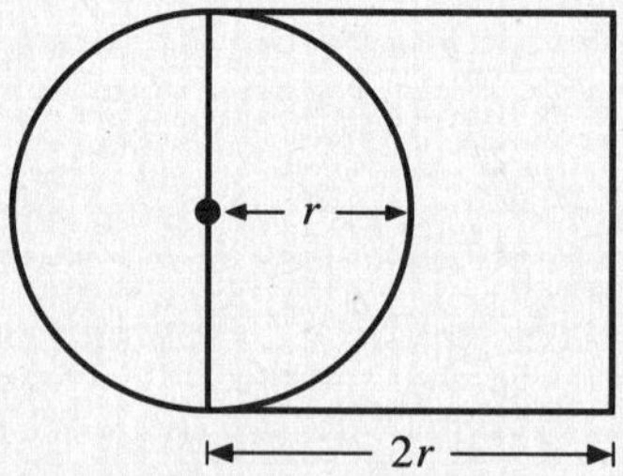

9. Find $P(A \cup B)$ when **9.** ________

$P(A) = 0.71$; $P(B) = 0.36$; $P(A \cap B) = 0.23$.

10. If $P(A) = 0.72$, what is $P(A')$? **10.** ________

11. ***Probability of Snow*** The probability for snow today is 0.6. The probability for snow tomorrow is 0.4. The probability for snow both days is 0.1. Find the probability that it will snow at least one of the two days. **11.** ________

Name ________________

12. A and B are independent events.

    $P(A) = 0.6$

    $P(B) = 0.8$

    Find $P(A \cap B)$.

12. ________________

13. ***Boys or Girls?*** Consider all families that have five children. If one family is chosen at random, what is the probability that the five children are all boys?

13. ________________

14. ***Marbles in a Jar*** A jar contains 21 green marbles and 30 yellow marbles. One marble is withdrawn and the color noted. It is then returned to the jar, mixed in, and another marble is withdrawn. Find the probability both marbles are green.

14. ________________

15. ***Expected Value*** Use the table below to find the expected value.

Payoff	\$20	\$15	\$10	\$12	\$16
Probability	0.10	0.25	0.33	0.20	0.12

15. ________________

16. ***Fair Game*** A single six-sided die is tossed and the number showing is paid in dollars (\$1 for a 1, \$2 for a 2, etc.). Find a fair price to pay to play this game.

16. ________________

Chapter 15 Test Form B

Name ____________________

(Page 1 of 3 pages)

Date ____________________

1. ***Choosing a Committee*** A four-person committee is chosen at random from a group of 16 people. How many different committees are possible? 1. ____________

2. ***Permuting Letters*** How many ways can the letters in the word HONEST be permuted? 2. ____________

3. ***Drawing Cards*** Three cards are drawn from a standard deck of playing cards. How many different such combinations are possible containing exactly two aces? 3. ____________

4. How many different three-digit numbers are possible if none of the digits are the same? 4. ____________

5. ***Drawing One Card*** A card is drawn from a standard deck of playing cards. Find the probability that it is not an ace or a heart. 5. ____________

6. ***Tossing Dice*** A pair of six-sided dice is tossed. The dots are added. In how many ways is it possible for the sum to total 6? 6. ____________

7. *Assigned Seats* You, your friend, and three others are assigned seats at random in a row of 5 chairs. How many ways can this be done and what is the probability that you and your friend will be seated next to each other?

7. ____________________

8. *Geometry* Half of a circle is inside a square and half is outside, as shown. If a point is selected at random inside the square, find the probability that the point is not inside the circle.

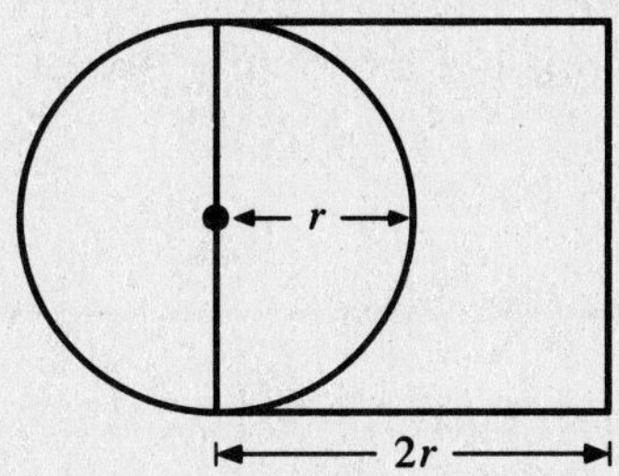

8. ____________________

9. Find $P(A \cup B)$ when

$$P(A) = 0.63; \; P(B) = 0.25; \; P(A \cap B) = 0.12.$$

9. ____________________

10. If $P(A) = 0.47$, what is $P(A')$?

10. ____________________

11. *Probability of Snow* The probability for snow today is 0.4. The probability for snow tomorrow is 0.35. The probability for snow both days is 0.17. Find the probability that it will snow at least one of the two days.

11. ____________________

12. A and B are independent events. 12. ____________

$P(A) = 0.7$

$P(B) = 0.32$

Find $P(A \cap B)$.

13. ***Boys or Girls?*** Consider all families that have five children. If one family is chosen at random, what is the probability that at least one of the children is a boy? 13. ____________

14. ***Marbles in a Jar*** A jar contains 21 green marbles and 30 yellow marbles. One marble is withdrawn and the color noted. It is then returned to the jar, mixed in, and another marble is withdrawn. Find the probability that both marbles are yellow. 14. ____________

15. ***Expected Value*** Use the table below to find the expected value. 15. ____________

Payoff	\$25	\$18	\$10	\$25	\$30
Probability	0.15	0.22	0.38	0.15	0.10

16. ***Fair Game*** Eight balls numbered 1 through 8 are placed in a container, thoroughly mixed, and one is withdrawn. The number on the ball is paid in dollars (\$1 for 1, \$2 for 2, etc.). Find a fair price to play this game. 16. ____________

Name ______________________

Date ______________________

1. ***3-Card Hands*** How many different 3-card hands can be drawn from a standard deck of 52 playing cards? 1. ____________

2. ***Permuting Letters*** How many recognizably different ways can the letters in the word SWEET be permuted? 2. ____________

3. ***Drawing Cards*** Three cards are drawn from a standard deck of playing cards. In how many ways is it possible to get a combination with exactly one ace and one king? 3. ____________

4. How many different three-digit numbers are possible where two *adjacent* digits are the same and the third digit is different? (Example: 110 and 011 are acceptable but 101 is not.) 4. ____________

5. ***Drawing One Card*** A card is drawn from a standard deck of playing cards. Find the probability that it is not a face card (**J**, **Q**, or **K**) or an ace. 5. ____________

6. ***Tossing Dice*** A pair of six-sided dice are tossed. The dots are added. In how many ways is it possible for the sum to total 7? Find the probability that the sum will total 7. 6. ____________

7. ***Assigning Seats*** You, your friend, and six others are assigned seats at random in a row of 8 chairs. How many ways can this be done and what is the probability that you and your friend will be seated next to each other?

7. ____________

8. ***Geometry*** A square is inscribed inside a circle as shown. If a point is chosen at random inside the circle, find the probability that the point is also inside the square.

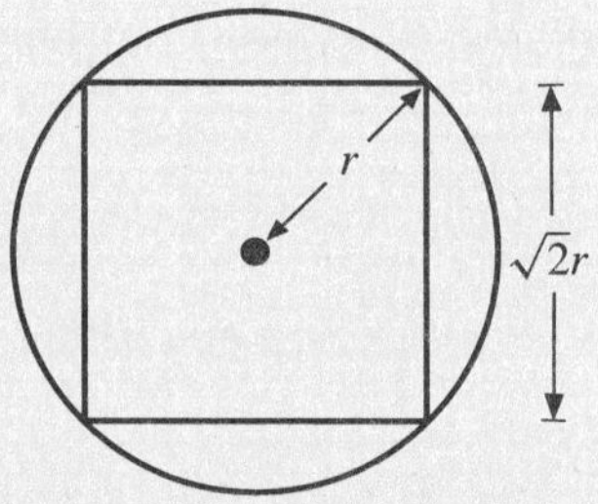

8. ____________

9. Find $P(A \cap B)$ when

$$P(A) = 0.54;\ \ P(B) = 0.42;\ \ P(A \cup B) = 0.85.$$

9. ____________

10. If $P(A) = 0.87$, what is $P(A')$?

10. ____________

11. ***Probability of Snow*** The probability for snow today is 0.48. The probability for snow tomorrow is 0.27. The probability for snow both days is 0.15. Find the probability that it will snow at least one of the two days.

11. ____________

Name ______________________

12. A and B are independent events.

$P(A) = \frac{3}{5}$

$P(B) = \frac{5}{6}$

Find $P(A \cap B)$.

12. ______________

13. ***Boys or Girls?*** Consider all families that have five children. If one family is chosen at random, what is the probability that exactly one of the five children is a boy?

13. ______________

14. ***Marbles in a Jar*** A jar contains 21 green marbles and 30 yellow marbles. One marble is withdrawn and the color noted. It is then returned to the jar, mixed in, and another marble is withdrawn. Find the probability that one marble of each color is obtained.

14. ______________

15. ***Expected Value*** Use the table below to find the expected value.

15. ______________

Payoff	\$25	\$15	\$10	\$5	\$35
Probability	0.12	0.18	0.25	0.38	0.07

16. ***Fair Game*** Ten balls numbered 1 through 10 are placed in a container, thoroughly mixed, and one is withdrawn. The number on the ball is paid in dollars (\$1 for 1, \$2 for 2, etc.). Find a fair price to play this game.

16. ______________

Cumulative Test

Chapters 13–15

(Page 1 of 8 pages)

Name ______________________

Date ______________________

In 1–5, evaluate without using a calculator.

1. $\sec 45^\circ$ — 1. __________

2. $\cos 30^\circ$ — 2. __________

3. $\sin 60^\circ$ — 3. __________

4. Find $\sec\theta$ when θ is an acute angle, and $\cos\theta = \frac{3}{4}$. — 4. __________

5. Find $\sin\theta$ when θ is an acute angle and $\tan\theta = \frac{12}{5}$. — 5. __________

In problems 6–10, use a calculator. Round to three decimal places.

6. $\csc 31^\circ$ — 6. __________

7. $\tan 87^\circ$ — 7. __________

8. $\sin 54^\circ$ — 8. __________

9. $\sec 118^\circ$ — 9. __________

10. $\cos(-84^\circ)$ — 10. __________

 Name ______

11. Find the complement of 66°. 11. ______

12. Find the complement of $\frac{3\pi}{8}$ radians. 12. ______

13. Find the supplement of 154°. 13. ______

14. Convert 72° to radian measure. 14. ______

15. Convert $\frac{8\pi}{3}$ radians to degree measure. 15. ______

16. The point $(-2, 7)$ is on the terminal side of an angle θ. Find $\sec\theta$. 16. ______

17. Evaluate $\arccos\left(-\frac{\sqrt{3}}{2}\right)$. Give your result in degree measure. 17. ______

18. Evaluate $\arctan(-0.507)$. Give your answer in radian measure. 18. ______

19. ***Geometry*** Solve for x in the triangle shown.

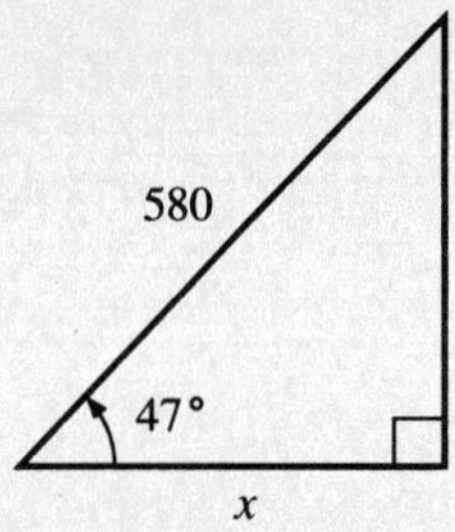

19. ____________________

20. ***Geometry*** Solve for y in the triangle shown.

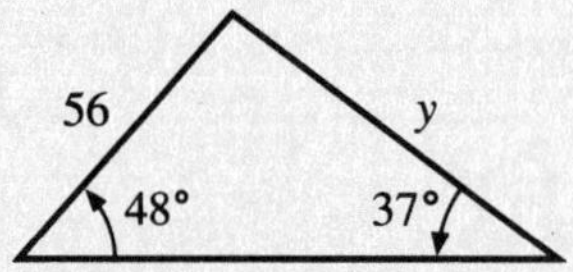

20. ____________________

21. ***Geometry*** Find the area of the triangle shown. Also find c.

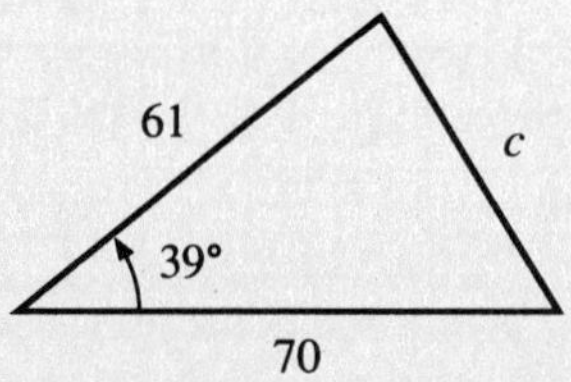

21. ____________________

22. Find the *amplitude* and *period* of the graph.

$$y = 8.14\cos(0.2x)$$

22. ____________________

23. Write an equation of the form $y = a \sin bx$ using

$a > 0$
$b > 0$
Amplitude: 7
Period: $\frac{2\pi}{3}$.

23. ____________________

 Name ______________________

24. Sketch one cycle of the graph of the function.

$$f(x) = 3\sin\frac{3\pi}{2}x$$

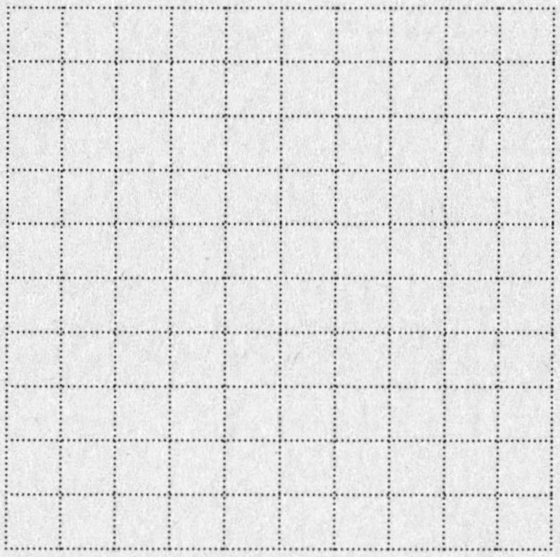

24. *Use graph at left.*

25. The graph of $y = \frac{2}{3}\cos 0.3x$ is drawn and then shifted two units to the right. Write the equation of the resulting graph.

25. ______________________

26. Which graph oscillates most frequently?

$$y = \tfrac{1}{2}\sin 4x \quad \text{or} \quad y = 2\sin\frac{x}{4}$$

26. ______________________

27. Sketch one cycle of the graph of the function.

$$y = 2 - 4\sin(\pi x - 2\pi)$$

27. *Use graph at left.*

28. Simplify the expression.

$$\csc x \tan(-x)$$

28. ______________________

29. Simplify the expression. **29.** ____________

$\cot x \cos(\frac{\pi}{2} - x)$

30. Verify the identity. **30.** ____________

$\cos^4 x - \sin^4 x = \cos^2 x - \sin^2 x$

31. Solve the equation for $0 \le x \le 2\pi$. **31.** ____________

$\tan^2 x - 3 = 0$

32. Simplify the expression. Do not evaluate it. **32.** ____________

$\cos 80° \cos 25° + \sin 80° \sin 25°$

33. Find the *exact* value of the expression. **33.** ____________

$\tan 75°$

34. Find the intersection points of the two graphs in the interval $0 \le x \le 2\pi$.

$y = 2\sin^2 x;\ y = \cos^2 x - \sin^2 x$

34. ______

35. Find the *exact* value of $\sin 15°$.

35. ______

36. Solve the equation for $0 \le x \le 2\pi$.

$\cos 2x - \sin x = 0$

36. ______

37. ***Permuting Letters*** How many ways can the letters in the word PERMUTE be permuted so the results are recognizably different?

37. ______

38. ***Forming Committees*** A three-person committee is to be formed from a group of 17 people. How many different committees are possible?

38. ______

39. ***Drawing Cards*** Three cards are drawn from a standard deck of playing cards. How many different such combinations that contain one red card and two black cards are possible? 39. ____________

40. ***Tossing a Coin*** A coin is tossed ten times and heads (H) or tails (T) is recorded as the result in each case. How many different outcomes are possible for the sequence of 10 tosses? 40. ____________

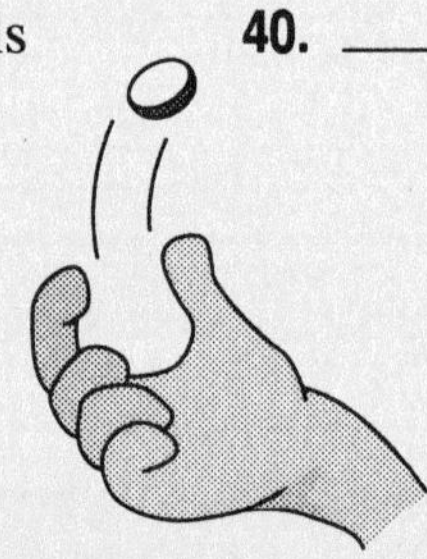

41. In Problem 40, in how many ways could the outcome produce 5 H and 5 T? 41. ____________

42. ***Assigning Seats*** You, your friend, and five others are assigned seats at random in a row of 7 chairs. How many ways can this be done and what is the probability that you and your friend will be seated next to each other? 42. ____________

43. If $P(A) = 0.457$, what is $P(A')$? 43. ____________

44. Find $P(A \cup B)$ when
$P(A) = 0.69$; $P(B) = 0.27$; $P(A \cap B) = 0.14$.

44. ________

45. ***Probability of Rain*** The probability for rain today is 0.70. The probability for rain tomorrow is 0.50. The probability for rain both days is 0.25. Find the probability that it will rain at least one of the two days.

45. ________

46. A and B are independent events. Find $P(A \cap B)$ when
$P(A) = 0.75$ and
$P(B) = 0.60$.

46. ________

47. ***Expected Value*** Use the table to find the expected value.

47. ________

Payoff	\$40	\$22	\$10	\$20
Probability	0.05	0.25	0.47	0.23

48. ***Fair Game*** Two coins are tossed. If both are tails, (TT), you get nothing. If one is tails, (TH or HT), you get \$10. If both are heads, (HH), you get \$40. Find a fair price to play this game.

48. ________

Strategies for Taking College Entrance Tests

Do your schoolwork.

College entrance mathematics tests are based on the concepts and skills taught in high school mathematics courses. The best way to prepare for a college entrance test is to keep up with your regular studies.

Learn about the exam.

You will be more prepared for taking the test if you are familiar with the directions and the format of the questions. This will reduce your anxiety and allow you to concentrate on the test questions and improve your chances of getting a good score.

While taking this algebra course, use the *Practice for College Entrance Tests* that are correlated with each chapter. These sample questions cover the topics you are currently learning.

Read the free booklets provided with the registration packets for the Scholastic Aptitude Test (SAT) or the American College Testing Assessment (ACT). These booklets are called *Taking the SAT* and *Preparing for the ACT Assessment.* They contain current information about the test you will take and a sample test. Many test preparation books, materials, and programs are also available that provide practice in taking college entrance tests. If you want additional practice, you can obtain materials from your school or public library. Your teacher and the guidance department at your school will also be able to provide information about test preparation programs and materials.

Study the directions in each section of the sample exam. This will save you time during the test since the directions printed in the registration booklets are the same as the directions that will appear on the test that you take.

Pace yourself.

Don't get bogged down in one question. If the question seems too difficult, skip it and go back after you have answered the other questions. When you go back to tackle a more difficult question, you may get an insight that you didn't have when you first read the question.

Don't rush through the test. Be sure you read each question and the possible answers. Decide how you want to check your pace. Don't, however, check the clock after answering each question. This tends to use time and increase anxiety. You might consider checking the time after every ten questions. Because the easier questions tend to be at the beginning of the test, you should try to allow more time for the later questions. Not all test taking sites have clocks on the wall, so you may want to bring a watch. Don't panic if you do not have time to answer all the questions. It is possible to receive a high score on a test without answering every question.

Make notes in the booklet.

When working on a question, make notes in the test booklet of what you do know. Label graphs and geometric figures with known and deduced information. If you decide to skip the question, this information will help refresh your memory when you return to it. To help avoid marking responses incorrectly, you should also circle the answer to each question in the test booklet.

Don't make wild guesses.

On some college entrance tests a fraction of a point is deducted for each wrong answer. If you don't know the answer to a question, try to eliminate one or two of the answer choices to increase the probability of choosing the correct answer. Then use your best judgement to choose the answer to the question from the remaining choices.

Check your work.

Once you have answered all the questions to a section in the test, use the remaining time to check the work you have done. This is the only time you will be allowed to review your notes and make changes. You are not permitted to return to a section of the test after the time has expired.

To check your work, first make sure that you have answered all the questions in the section and have indicated only one answer for each question. Look for and remove any stray pencil marks that could affect the scoring of the test.

Next, review each problem and check any calculations you have made. This checking will be easier if you made clear notes in the test booklet when you originally worked the problem.

Know your calculator.

Some college entrance tests allow the optional use of calculators. Remember that a calculator is only a tool—it cannot answer the questions for you. The calculator that you bring to the test should be one that you have used. Be sure that it uses the order of operations that you expect. For instance, if you enter $4 + 6/2 - 1 =$, does your calculator display 6 or 4? If it displays 6, then it is using the correct order of operations, $4 + (6/2) - 1 = 6$. If it displays 4, then it is using the following order of operations: $[(4+6)/2]-1 = 4$. The important thing is that you must know how the calculator you are using works. When you use a calculator, be careful to press the correct keys.

Relax the night before.

On the night before the exam, gather together any items you will need for the test: pencils with erasers, your admission ticket, identification, directions to the test center, glasses for reading, a watch to pace yourself, and (if allowed) a calculator with *fresh batteries*.

Don't try to cram the night before the test. Do something that you find relaxing. Get a good night's sleep and eat breakfast in the morning.

Instructions for Recording Student-Generated Responses

1994 SAT Test

Starting in the spring of 1994, the Scholastic Aptitude Test will include 10 student-produced response questions. The student will solve the problem and then write the answer in a grid. Below are some sample questions and answers to demonstrate how to use the grid. In each case, notice that you write your answer at the top of the grid. Next, fill in completely the corresponding ellipses in the column. The answer may be an integer, a decimal, or a fraction.

1.

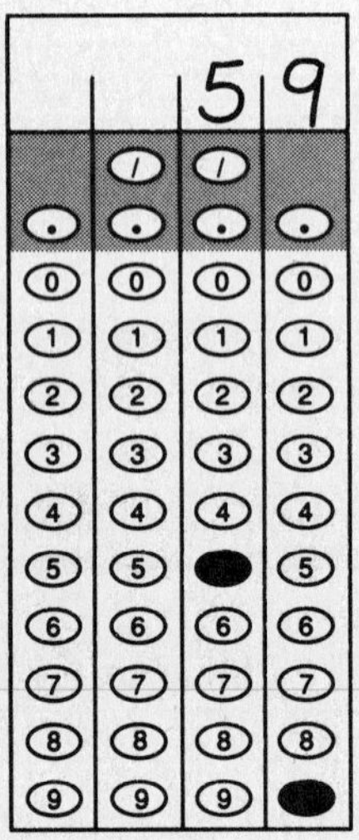

If the eighth number in the list is 47, and if each number after the first number in the list is 3 less than the number preceding it, what is the fourth number in the list?

Answer: 59

2.

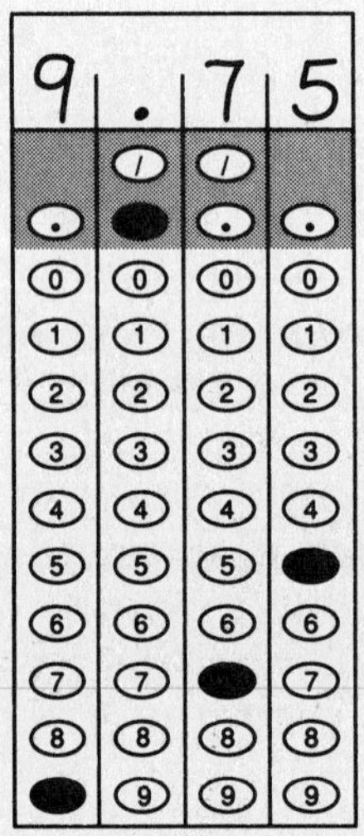

Apples are sold 8 for $1. At this rate if 2 apples are bought and paid for with a $10.00 bill, what is the correct change, in dollars? (Assume no tax.)

Answer: 9.75

3.

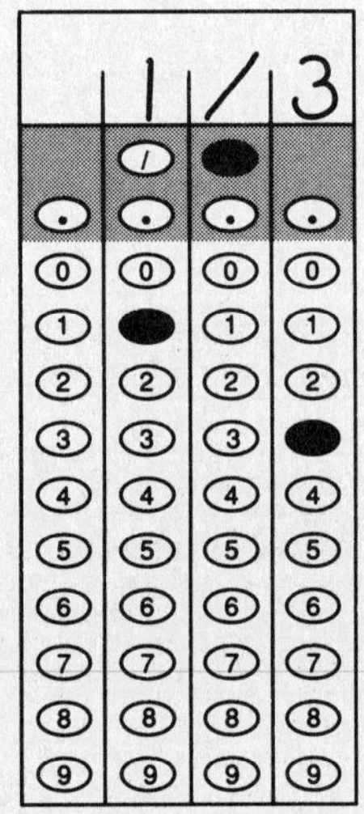

An oatmeal cookie recipe calls for 1 cup of raisins for every 3 cups of flour. If a small batch of cookies is desired using only 1 cup of flour, how many cups of raisins should be used?

Answer: $\frac{1}{3}$

Answer Form for Student-Generated Responses

Chapter ______ Name ____________________

1.

2.

3.

4.

5.

6.

7.

8.

9.

10.

11.

12.

Practice for College Entrance Tests
Chapter 1

Form A Name ______________________

Fill in the circle of the letter that corresponds to the correct answer.

1. The expression $(3 + 2) \cdot 5 - 4$ is equivalent to [?]

   A. 5 **B.** 16 **C.** 21 **D.** 26 **E.** 29

   1. Ⓐ Ⓑ Ⓒ Ⓓ Ⓔ

2. The fine for speeding is \$25 plus \$4 for each mile per hour over the speed limit. If a car is stopped for traveling 9 miles per hour over the speed limit, what is the amount of the speeding ticket?

   A. \$25 **B.** \$29 **C.** \$38 **D.** \$61 **E.** \$100

   2. Ⓐ Ⓑ Ⓒ Ⓓ Ⓔ

3. If $4x + 2 \geq 22$, then possible values for x are:

   I. 4 II. 5 III. 7.

   A. I only **B.** II only **C.** III only
   D. I and II **E.** II and III

   3. Ⓐ Ⓑ Ⓒ Ⓓ Ⓔ

4. Four times the sum of a number and ten may be written as which of the following?

   A. $4n + 10$ **B.** $4n - 10$ **C.** $\frac{n+10}{4}$
   D. $4(n + 10)$ **E.** $n + 4 \cdot 10$

   4. Ⓐ Ⓑ Ⓒ Ⓓ Ⓔ

5. Which equation has the solution points shown on the graph at the right?

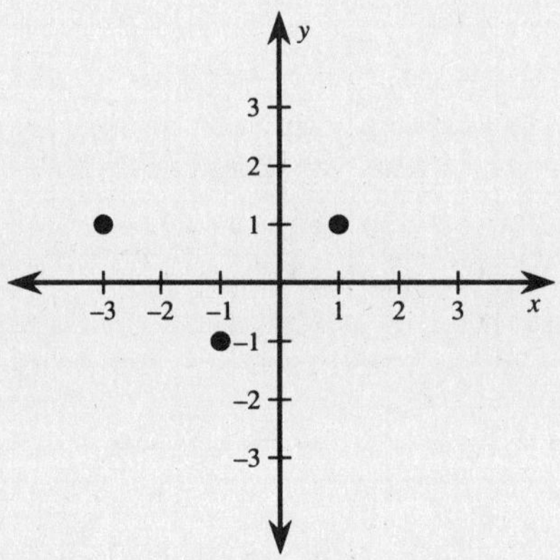

   A. $y = |x + 1| - 3$ **B.** $y = |x - 3| + 1$ **C.** $y = |x + 1| - 1$
   D. $y = |x + 3| - 1$ **E.** $y = |x - 1| + 1$

   5. Ⓐ Ⓑ Ⓒ Ⓓ Ⓔ

6. What is the sum of the 2 solutions for x in the equation $|x + 3| - 5 = -2$?

   A. -6 **B.** -4 **C.** 0 **D.** 2
   E. The equation has no solutions

   6. Ⓐ Ⓑ Ⓒ Ⓓ Ⓔ

7. A rectangle has sides of $(x + 3)$ feet and $(x + 7)$ feet. How many feet is the perimeter?

   A. $2x + 10$ **B.** $2x + 20$ **C.** $4x + 10$
   D. $4x + 20$ **E.** $x^2 + 21$

   7. Ⓐ Ⓑ Ⓒ Ⓓ Ⓔ

Practice for College Entrance Tests

Chapter 1

Form B **Name** ____________

If Column A is greater than Column B, fill in the circle for A.
If Column B is greater than Column A, fill in the circle for B.
If Column A is equal to Column B, fill in the circle for C.
If there is not enough information, fill in the circle for D.
Information centered between the columns concerns the quantities in both columns.

	Column A		Column B	
1.	$(2 \times 10^2) + (4 \times 10^4)$		$(2 \times 10^2) + (6 \times 10^3)$	1. Ⓐ Ⓑ Ⓒ Ⓓ
2.	2^4		4^2	2. Ⓐ Ⓑ Ⓒ Ⓓ
3.		$y = 3$		3. Ⓐ Ⓑ Ⓒ Ⓓ
	y^3		9	
4.		4 less than $5x$ is 16.		4. Ⓐ Ⓑ Ⓒ Ⓓ
	3		x	
5.		The perimeter of a rectangle A is equal to the perimeter of an equilateral triangle B.		5. Ⓐ Ⓑ Ⓒ Ⓓ
	Length of a side of A		Length of a side of B	
6.		a, b, and c are consecutive integers and $abc = 0$.		6. Ⓐ Ⓑ Ⓒ Ⓓ
	The greatest of a, b, and c		0	
7.		$a = 2b$		7. Ⓐ Ⓑ Ⓒ Ⓓ
	Area of a circle with a radius of $\frac{a}{2}$		Area of a circle with a radius of b	
8.		(see table below)		8. Ⓐ Ⓑ Ⓒ Ⓓ
	The cost of a compact disk and a cassette		The cost of 3 cassettes	

Item	Price
compact disk	\$12.00
cassette	\$5.00

Practice for College Entrance Tests
Chapter 2

Form A **Name** ____________________

Fill in the circle of the letter that corresponds to the correct answer.

1. Which of the following graphs represents the solution to $2x - 1 \leq 3$? **1.** Ⓐ Ⓑ Ⓒ Ⓓ Ⓔ

A.

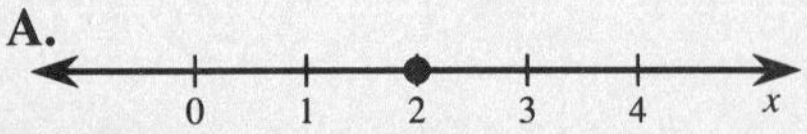

B.

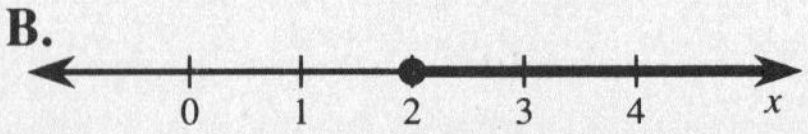

C.

D.

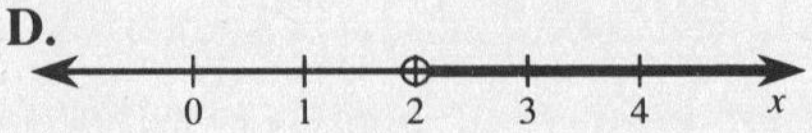

E.

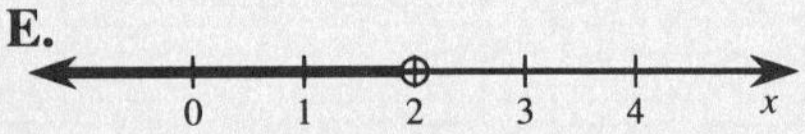

2. The Lotsa Toys Company has an increase in earnings as shown in the graph at the right. What is the company's average yearly increase? **2.** Ⓐ Ⓑ Ⓒ Ⓓ Ⓔ

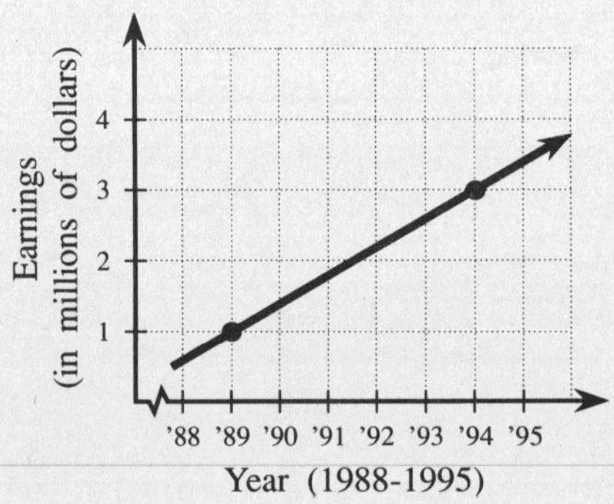

A. \$400,000 **B.** \$500,000 **C.** \$800,000 **D.** \$1,000,000
E. \$2,000,000

3. If $p + q = 7$, $q - r = 5$, and $q = -3$, then $r - p =$ [?] **3.** Ⓐ Ⓑ Ⓒ Ⓓ Ⓔ

A. -18 **B.** -2 **C.** 2 **D.** 4 **E.** 18

4. A savings account has an initial deposit of \$50. Each week thereafter \$10 is deposited into the account. Which equation represents the balance, b, after w weeks? **4.** Ⓐ Ⓑ Ⓒ Ⓓ Ⓔ

A. $b = 50 + 10w$ **B.** $b = 10 + 50w$ **C.** $b = 50 + 10 + w$
D. $w = 10 + 50b$ **E.** $w = 50 + 10b$

5. What is the slope, m, and the y-intercept, b, of the graph of the line $3x - 2y = 4$? **5.** Ⓐ Ⓑ Ⓒ Ⓓ Ⓔ

A. $m = \frac{2}{3}$; $b = -2$ **B.** $m = -\frac{3}{2}$; $b = 2$ **C.** $m = \frac{3}{2}$; $b = -2$
D. $m = -\frac{2}{3}$; $b = 2$ **E.** $m = \frac{3}{2}$; $b = 2$

6. What is the equation of the line with the same slope as $y = 3x - 1$ but with the same y-intercept as $4x - y = 2$? **6.** Ⓐ Ⓑ Ⓒ Ⓓ Ⓔ

A. $y = 3x - 2$ **B.** $y = 4x - 1$ **C.** $y = 3x + 2$
D. $y = -4x + 1$ **E.** $y = -x + 2$

Practice for College Entrance Tests
Chapter 2

Form B Name ______________________

Solve the problem and write your answer at the top of the grid on the separate answer sheet. Then fill in the corresponding ellipses in the answer grid. (Your answer can be an integer, a decimal number, or a fraction.)

In 1 and 2 refer to the graph below.

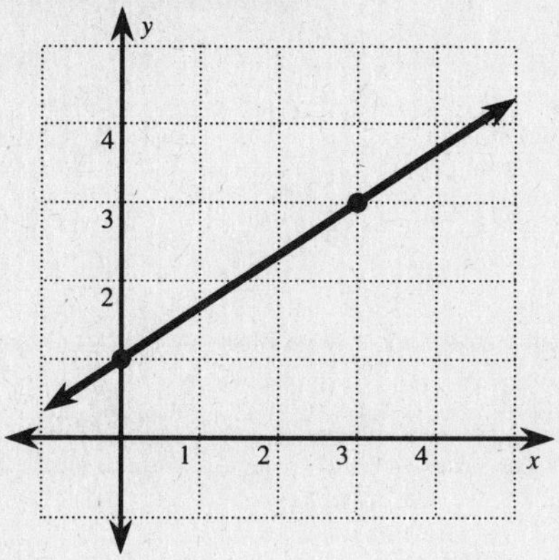

1. What is the slope of the line?

2. What is the y-intercept of the line?

3. If the ratio of x to y is $\frac{3}{2}$ and the ratio of x to z is $\frac{7}{4}$, what is the ratio of y to z?

4. From 7:00 A.M. to 1:00 P.M., the temperature rose at a constant rate from 8°C to 20°C. What was the temperature (in °C) at 11:00 A.M.?

5. What is the y-intercept of the line $4x - y = -4$?

6. What is the x-intercept of the line $3x + 4y = 7$?

7. For what values of x will the graph of the line $y = |x - 5| - 4$ have negative y values?

8. Sandra earns \$1 an hour for babysitting and \$0.75 an hour for running errands for neighbors. She wants to earn \$80 to buy a new bike. If Sandra babysits for 54.5 hours, how many hours must she run errands to have enough money for the new bike?

Practice for College Entrance Tests Chapter 3

Form A Name ______________________

Fill in the circle of the letter that corresponds to the correct answer.

1. If a is 4 more than the quotient of b and 3, and if $b = 2$, then what is the value of a?

A. $-3\frac{1}{3}$ B. $\frac{1}{6}$ C. 2 D. $4\frac{2}{3}$ E. 10

1. Ⓐ Ⓑ Ⓒ Ⓓ Ⓔ

2. If $3x - 4y = 7$, then $2y =$ [?]

A. $-\frac{3}{2}x - \frac{7}{2}$ B. $-\frac{3}{2}x + \frac{7}{2}$ C. $-\frac{3}{4}x - \frac{7}{4}$
D. $\frac{3}{4}x - \frac{7}{4}$ E. $\frac{3}{2}x - \frac{7}{2}$

2. Ⓐ Ⓑ Ⓒ Ⓓ Ⓔ

3. If $4p + q = 13$ and $p - q = 2$, then $pq =$ [?]

A. 2 B. 3 C. 4 D. 15 E. 26

3. Ⓐ Ⓑ Ⓒ Ⓓ Ⓔ

4. If $2x - 4y = 5$ and $3x - y = -5$ are equations which represent the sides $\overline{PQ}$ and $\overline{PS}$ of quadrilateral $PQRS$, what are the (x, y) coordinates of P?

A. $\left(\frac{25}{14}, \frac{15}{7}\right)$ B. $\left(\frac{3}{2}, 2\right)$ C. $\left(-\frac{4}{3}, 1\right)$
D. $\left(-\frac{3}{2}, \frac{1}{2}\right)$ E. $\left(-\frac{5}{2}, -\frac{5}{2}\right)$

4. Ⓐ Ⓑ Ⓒ Ⓓ Ⓔ

5. If $-2x + 3y = -12$ and $3x + 3y = 3$, then $x + y =$ [?]

A. $\frac{17}{5}$ B. 1 C. -1 D. -15 E. -19

5. Ⓐ Ⓑ Ⓒ Ⓓ Ⓔ

6. Which of the following systems of equations represents the graph to the right?

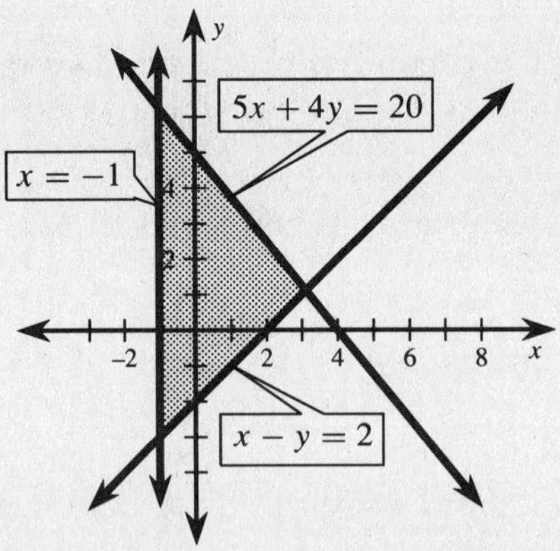

6. Ⓐ Ⓑ Ⓒ Ⓓ Ⓔ

A. $\begin{cases} 5x + 4y \leq 20 \\ x - y \leq 2 \\ x \geq -1 \end{cases}$ B. $\begin{cases} 5x + 4y \leq 20 \\ x - y \leq 2 \\ x \leq -1 \end{cases}$ C. $\begin{cases} 5x + 4y \geq 20 \\ x - y \leq 2 \\ x \leq -1 \end{cases}$

D. $\begin{cases} 5x + 4y \geq 20 \\ x - y \geq 2 \\ x \geq -1 \end{cases}$ E. $\begin{cases} 5x + 4y \leq 20 \\ x - y \geq 2 \\ x \leq -1 \end{cases}$

Practice for College Entrance Tests

Chapter 3

Form B Name ____________________

If Column A is greater than Column B, fill in the circle for A.
If Column B is greater than Column A, fill in the circle for B.
If Column A is equal to Column B, fill in the circle for C.
If there is not enough information, fill in the circle for D.
Information centered between the columns concerns the quantities in both columns.

	Column A		Column B	
1.		$5x - 10y = 15$		**1.** Ⓐ Ⓑ Ⓒ Ⓓ
	5		$x - 2y$	
2.		$2x - 3y = -1$ $x = 2$		**2.** Ⓐ Ⓑ Ⓒ Ⓓ
	-2		y	
3.		$x = \frac{2}{3}y$ and $x = 4 + 2y$		**3.** Ⓐ Ⓑ Ⓒ Ⓓ
	-3		y	
4.		$\begin{cases} 3x - 2y = 14 \\ 2x + 2y = 6 \end{cases}$		**4.** Ⓐ Ⓑ Ⓒ Ⓓ
	$x - y$		5	
5.		$\begin{cases} x - y < 2 \\ x + y < 2 \end{cases}$		**5.** Ⓐ Ⓑ Ⓒ Ⓓ
	x		3	
6.		$a + 5 = b$ $ab \neq 0$		**6.** Ⓐ Ⓑ Ⓒ Ⓓ
	The average (arithmetic mean) of a, b, and 5		$a - b$	
7.		$3x - 2y = 8$ and $x + 2y = 0$		**7.** Ⓐ Ⓑ Ⓒ Ⓓ
	$x - y$		$x + y$	
8.		$2x + 3y = 5a$ and $-x + 2y = a$		**8.** Ⓐ Ⓑ Ⓒ Ⓓ
	x		y	

Practice for College Entrance Tests Chapter 4

Form A **Name** ____________________

Fill in the circle of the letter that corresponds to the correct answer.

1. If $3x - 2 < 10$, then $\boxed{?}$ 1. Ⓐ Ⓑ Ⓒ Ⓓ Ⓔ

A. $x \leq \frac{8}{3}$ B. $x < 4$ C. $x \leq 4$
D. $x > 4$ E. $x < \frac{10}{3}$

2. A rectangle has sides of 3 feet and 7 feet. How many feet is its perimeter? 2. Ⓐ Ⓑ Ⓒ Ⓓ Ⓔ

A. 10 B. 20 C. 13 D. 17 E. 21

3. What is the value of $3x^2 + 2y$ when $x = 1$ and $y = 4$? 3. Ⓐ Ⓑ Ⓒ Ⓓ Ⓔ

A. 5 B. 11 C. 14 D. 27 E. 56

4. If $x = \frac{2}{3}$ and $y = -2$, then $x^2 - 3y + \frac{2}{9} = \boxed{?}$ 4. Ⓐ Ⓑ Ⓒ Ⓓ Ⓔ

A. $-5\frac{1}{3}$ B. $-5\frac{1}{9}$ C. $2\frac{2}{9}$ D. $6\frac{4}{9}$ E. $6\frac{2}{3}$

5. In one day a hot dog stand sells 200 hot dogs. If 75% of the hot dogs are ordered with catsup and of these 40% are ordered with mustard, how many hot dogs are ordered with catsup and mustard? 5. Ⓐ Ⓑ Ⓒ Ⓓ Ⓔ

A. 30 B. 40 C. 60 D. 75 E. 115

6. A car traveled at an average speed of 60 mph for 4 hours. On the return trip the car averaged 40 mph. What was the average speed, in miles per hour, for the entire trip? 6. Ⓐ Ⓑ Ⓒ Ⓓ Ⓔ

A. 48 B. 49 C. 50 D. 51 E. 52

7. The sum of two numbers is 20 and the ratio of the smaller number to the larger number is 1:3. What is the smaller number? 7. Ⓐ Ⓑ Ⓒ Ⓓ Ⓔ

A. 2 B. 3 C. 4 D. 5 E. 6

8. The correct addition problem is shown below. If M and N are 2 different non-zero digits, then what is the value of N? 8. Ⓐ Ⓑ Ⓒ Ⓓ Ⓔ

$$\begin{array}{r} 5M \\ +M1 \\ \hline 1N9 \end{array}$$

A. 1 B. 3 C. 4 D. 6 E. 8

9. If the sum of two consecutive odd integers is 48, what is their product? 9. Ⓐ Ⓑ Ⓒ Ⓓ Ⓔ

A. 96 B. 506 C. 575 D. 650 E. 675

Form B Name ____________________

Solve the problem and write your answer at the top of the grid on the separate answer sheet. Then fill in the corresponding ellipses in the answer grid. (Your answer can be an integer, a decimal number, or a fraction.)

1. What is the integer between $\frac{17}{5}$ and $\frac{31}{7}$?

2. Manuel's grades for the first five history exams this semester were 100, 87, 92, 93, and 89. What must his next test grade be in order to have an average of 92 this semester?

3. What is the decimal representation of $2 + \frac{3}{10} + \frac{7}{100}$?

4. If $a = 3m + n$ and $b = m - 2n$, what is $a - b$ when $m = 1$ and $n = 2$?

5. What is the product of $\frac{2}{5}$ and $\frac{75}{4}$?

6. A family is moving from city A to city B. The moving van leaves city A at 10 a.m. Traveling at an average speed of 45 miles per hour, the moving van arrives in city B at 4 p.m. If the family travels by car at an average speed of 54 miles per hour and arrives in city B at the same time as the moving van, how many hours after the moving van left city A did the family leave?

7. If $-\dfrac{b+2}{3} + 2 = -1$ then $b = \boxed{\ ?\ }$

8. What is the value of y in the solution of the system below?

$$\begin{cases} 3x + 2y = 8 \\ -x + 2y = 4 \end{cases}$$

9. There are 60 meters of fence to surround a rectangular garden. If all of the fence is to be used and the length is between 18 and 20 meters, grid in a possible value for the width of the garden.

10. Two hundred cars drove into the state park in one hour. The number of persons in each car varied. The tally is in the table below. What was the average number of persons per car for the one hour?

Number of persons per car	1	2	3	4	5
Number of cars	20	40	50	60	30

Practice for College Entrance Tests
Chapter 5

Form A Name ______________________

Fill in the circle of the letter that corresponds to the correct answer.

1. What is the value of $3x^2 - 2x + 6$ when $x = -2$ **1.** Ⓐ Ⓑ Ⓒ Ⓓ Ⓔ

A. -10 **B.** -2 **C.** 4 **D.** 14 **E.** 22

2. If $x = 5$ is a solution to $ax^2 - 13x - 10 = 0$, what is the value of a? **2.** Ⓐ Ⓑ Ⓒ Ⓓ Ⓔ

A. 1 **B.** 2 **C.** 3 **D.** 4 **E.** 5

3. One leg of a right triangle has a length of 9 cm and the other leg has a length of 40 cm. What is the length of the hypotenuse? **3.** Ⓐ Ⓑ Ⓒ Ⓓ Ⓔ

A. 31 **B.** 37 **C.** 39 **D.** 41 **E.** 49

4. If $p > 0$, what are the roots of $x^2 + 5px + 4p^2 = 0$? **4.** Ⓐ Ⓑ Ⓒ Ⓓ Ⓔ

A. $-2p$ and $-3p$ **B.** $-p$ and $-4p$ **C.** $-p$ and $6p$
D. p and $4p$ **E.** $2p + 3p$

5. You want to use a 12-foot ladder to clean the windows of your house. If the base of the ladder must be at least 3 feet from the base of the house, approximately how far up the side of the house can the ladder reach? **5.** Ⓐ Ⓑ Ⓒ Ⓓ Ⓔ

A. $\approx$ 9 feet **B.** $\approx$ 9.6 feet **C.** $\approx$ 10.6 feet
D. $\approx$ 11.6 feet **E.** $\approx$ 15 feet

6. Which set contains both solutions to $x^2 - 2x - 8 = 0$? **6.** Ⓐ Ⓑ Ⓒ Ⓓ Ⓔ

A. $\{-8, -6, -4, -2\}$ **B.** $\{-8, -6, -2, 4\}$ **C.** $\{-6, -4, 2, 8\}$
D. $\{-6, 2, 4, 8\}$ **E.** $\{2, 4, 6, 8\}$

7. If p and q are positive numbers and $p^2 + q^2 = 29$ and $p^2 - q^2 = 21$, what does q equal? **7.** Ⓐ Ⓑ Ⓒ Ⓓ Ⓔ

A. 25 **B.** 8 **C.** 5 **D.** 4 **E.** 2

8. Which of the following expressions represent the solutions for $x^2 + 3x - 10 > 0$? **8.** Ⓐ Ⓑ Ⓒ Ⓓ Ⓔ

A. $x < -5$ or $x > 2$ **B.** $-5 < x < 2$ **C.** $x = -5$ and $x = 2$
D. $x < -2$ or $x > 5$ **E.** $-2 < x < 5$

9. Which point is not a solution to $y \le \frac{1}{3}x^2 + 2x - 1$? **9.** Ⓐ Ⓑ Ⓒ Ⓓ Ⓔ

A. $(-9, 0)$ **B.** $(-3, -5)$ **C.** $(0, 3)$
D. $(3, 6)$ **E.** $(6, 0)$

Practice for College Entrance Tests

Chapter 5

Form B **Name** ______________________

If Column A is greater than Column B, fill in the circle for A.
If Column B is greater than Column A, fill in the circle for B.
If Column A is equal to Column B, fill in the circle for C.
If there is not enough information, fill in the circle for D.
Information centered between the columns concerns the quantities in both columns.

	Column A		Column B	
1.	$15^2 - 12^2$		9^2	**1.** Ⓐ Ⓑ Ⓒ Ⓓ
2.		$m \geq 1$ and $m^2 = n$		**2.** Ⓐ Ⓑ Ⓒ Ⓓ
	n		1	
3.		The formula for the volume of a pyramid is $\frac{1}{3}hs^2$, where h is the height and s is the length of the side of its base.		**3.** Ⓐ Ⓑ Ⓒ Ⓓ
	The length of the side of a pyramid with a volume of 25 and height of 3		The length of the side of a pyramid with a volume of 15 and height of 5	
4.	$m^2 - 3$		$m^2 - 1$	**4.** Ⓐ Ⓑ Ⓒ Ⓓ
5.		$x^2 - 3x - 18 = 0$		**5.** Ⓐ Ⓑ Ⓒ Ⓓ
	x		3	
6.		$y \leq x^2 - 3x + 4$		**6.** Ⓐ Ⓑ Ⓒ Ⓓ
	The greatest possible value of y when $x = -2$		The greatest possible value of y when $x = 2$	
7.		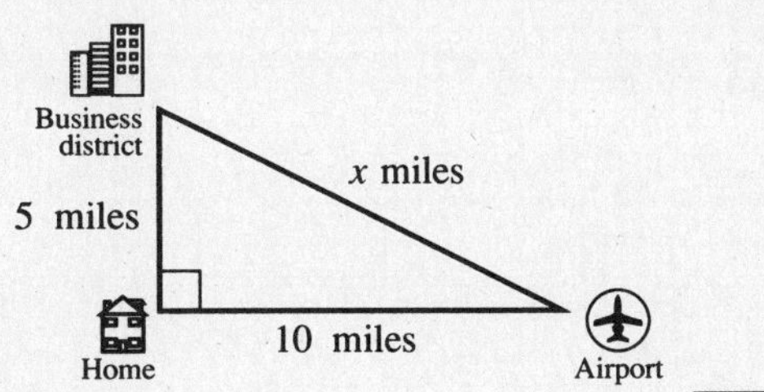		**7.** Ⓐ Ⓑ Ⓒ Ⓓ
	The distance, x, from the airport to the business district		$\sqrt{130}$	

Practice for College Entrance Tests Chapter 6

Form A Name ______________________

Fill in the circle of the letter that corresponds to the correct answer.

1. What is the value of the function $f(x) = 4x - 2$ when $x = 2$? **1.** Ⓐ Ⓑ Ⓒ Ⓓ Ⓔ

A. 2 **B.** 4 **C.** 6 **D.** 8 **E.** 10

2. If $\boxed{x} = x^2 - 2x + 1$, what is the value of $\boxed{4}$? **2.** Ⓐ Ⓑ Ⓒ Ⓓ Ⓔ

A. -3 **B.** 3 **C.** 5 **D.** 7 **E.** 9

3. Which of the following represents $f(x) = 3x^2 + 2x - 6$ in completed square form? **3.** Ⓐ Ⓑ Ⓒ Ⓓ Ⓔ

A. $f(x) = 3(x + \frac{1}{3})^2 - 6$ **B.** $f(x) = 3(x + \frac{1}{3})^2 - \frac{19}{3}$
C. $f(x) = (x + \frac{1}{3})^2 - \frac{19}{3}$ **D.** $f(x) = (x + \frac{1}{3})^2 - \frac{19}{9}$
E. $f(x) = (x + \frac{1}{3})^2 - 2$

In 4 and 5, use the following information.

Pat's English test scores for this semester were: 79, 88, 84, 91, 88, *and* 86.

4. What was Pat's mean score for the semester? **4.** Ⓐ Ⓑ Ⓒ Ⓓ Ⓔ

A. 85 **B.** 86 **C.** 87 **D.** 88 **E.** 89

5. What was the mode of Pat's scores? **5.** Ⓐ Ⓑ Ⓒ Ⓓ Ⓔ

A. 79 **B.** 84 **C.** 86 **D.** 88 **E.** 91

In 6 and 7, use the definition $\boxed{x} = x^2 + 3x - 7$.

6. What is the value of $\boxed{-2} + \boxed{1}$? **6.** Ⓐ Ⓑ Ⓒ Ⓓ Ⓔ

A. -12 **B.** -9 **C.** -6 **D.** -3 **E.** 0

7. If $\boxed{x} = -3$, what are the possible values of x? **7.** Ⓐ Ⓑ Ⓒ Ⓓ Ⓔ

I. -5 II. -4 III. 1 IV. 2

A. I. only **B.** II. only **C.** IV. only
D. I. and IV. only **E.** II. and III. only

8. If $f(x) = 2x^2 - 3x - 4$ and $g(x) = x^2 + 2x + 1$, what does $f(x) - g(x)$ equal? **8.** Ⓐ Ⓑ Ⓒ Ⓓ Ⓔ

A. $3x^2 - x - 3$ **B.** $x^2 - x - 3$ **C.** $x^2 - 5x - 5$
D. $-x^2 + 5x + 5$ **E.** $-x^2 - x - 3$

Practice for College Entrance Tests Chapter 6

Form B Name ____________________

Solve the problem and write your answer at the top of the grid on the separate answer sheet. Then fill in the corresponding ellipses in the answer grid. (Your answer can be an integer, a decimal number, or a fraction.)

1. Grid in a value of x such that the value of the function $f(x) = 3x+2$ has a value between 5 and 17.

2. When the function $f(x) = x^2 - 7x + 12$ is equal to 0 there are two possible values for x. What is one of those values?

3. The function $g(x) = ax^2 + 19x - 5$ is equal to 0 when $x = -5$. What is the value of a?

4. What is a possible number in the domain of the function, $f(x) = \sqrt{x-5}$?

5. What is a number that is not in the domain of the function $g(x) = \dfrac{2}{x-7}$?

6. What is the value of the function $P(x) = \dfrac{3}{x^2+3}$ when $x = 5$?

In 7 and 8, use the following definition.

$\boxed{x} = x^2 + 3$

7. What is the sum of $\boxed{2}$ and $\boxed{3}$?

8. If $\boxed{1} \cdot \boxed{2} = \boxed{x}$ what is the value of x?

In 9–11, use the table.

Hours Studied

	Mon	Tues	Wed	Thur	Fri
Pat	$1\frac{1}{2}$	1	$2\frac{1}{2}$	$1\frac{1}{2}$	1
Juan	2	$1\frac{1}{2}$	2	$1\frac{1}{2}$	1
Chuck	1	2	$1\frac{1}{2}$	2	$1\frac{1}{2}$

9. What is the average number of hours per day that Pat studied after school?

10. What is the mode of the number of hours studied per day?

Practice for College Entrance Tests
Chapter 7

Form A Name ______________________

Fill in the circle of the letter that corresponds to the correct answer.

1. Which of the following is equivalent to $m^6 \cdot m^2$? 1. Ⓐ Ⓑ Ⓒ Ⓓ Ⓔ

   A. m^2 **B.** m^3 **C.** m^4 **D.** m^8 **E.** m^{12}

2. Which of the following is not equivalent to $2^3 \times 3^2$? 2. Ⓐ Ⓑ Ⓒ Ⓓ Ⓔ

   I. 2×3^3 II. 2×6^2 III. 72

   A. I. only **B.** II. only **C.** III. only
   D. I. and II. **E.** II. and III.

3. What is the average (arithmetic mean) of 2^2, 5^2, and 6^2? 3. Ⓐ Ⓑ Ⓒ Ⓓ Ⓔ

   A. $4\frac{1}{3}$ **B.** $10\frac{5}{6}$ **C.** $18\frac{7}{9}$ **D.** $21\frac{2}{3}$ **E.** $32\frac{1}{2}$

4. Which of the following is equivalent to $\frac{a^2b^{-1}c^4}{b^2} \cdot \frac{a^3c^2}{b^4c^3}$? 4. Ⓐ Ⓑ Ⓒ Ⓓ Ⓔ

   A. $\frac{a^6c^5}{b^9}$ **B.** $\frac{a^6c^3}{b^7}$ **C.** $\frac{a^5c^4}{b^7}$
   D. $\frac{a^5c^3}{b^7}$ **E.** $\frac{a^5c^3}{b^5}$

5. $\left(\sqrt{3} - \sqrt{5}\right)\left(\sqrt{5} - \sqrt{3}\right) = \boxed{?}$ 5. Ⓐ Ⓑ Ⓒ Ⓓ Ⓔ

   A. -8 **B.** $2 - 2\sqrt{15}$ **C.** -2 **D.** $2\sqrt{15} - 8$ **E.** 2

6. If $\frac{7^3 - 7^2}{6} = 7^x$, what is the value of x? 6. Ⓐ Ⓑ Ⓒ Ⓓ Ⓔ

   A. 0 **B.** 1 **C.** 2 **D.** 3 **E.** 4

7. $(2.2 \times 10^4) \div (5.0 \times 10^6) = \boxed{?}$ 7. Ⓐ Ⓑ Ⓒ Ⓓ Ⓔ

   A. 4.4×10^{-3} **B.** 1.1×10^{-1} **C.** 4.4×10^1
   D. 4.4×10^9 **E.** 1.1×10^{11}

8. A line segment has endpoints of $(-4, 2)$ and $(2, -3)$ in the (x, y) coordinate plane. What is the length of this line segment? 8. Ⓐ Ⓑ Ⓒ Ⓓ Ⓔ

   A. $\sqrt{61}$ **B.** 7 **C.** $\sqrt{11}$ **D.** $\sqrt{5}$ **E.** $\sqrt{3}$

9. The distance between points $(3, -4)$ and $(7, y)$ is $2\sqrt{13}$. What are the possible values for y? 9. Ⓐ Ⓑ Ⓒ Ⓓ Ⓔ

   A. -10 only **B.** -2 only **C.** 2 only
   D. -10 and -2 **E.** -10 and 2

Practice for College Entrance Tests

Chapter 7

Form B

Name ______________________

If Column A is greater than Column B, fill in the circle for A.
If Column B is greater than Column A, fill in the circle for B.
If Column A is equal to Column B, fill in the circle for C.
If there is not enough information, fill in the circle for D.
Information centered between the columns concerns the quantities in both columns.

	Column A		Column B	
1.	3^4		4^3	**1.** Ⓐ Ⓑ Ⓒ Ⓓ
2.	p^2q^4		p^3q^4	**2.** Ⓐ Ⓑ Ⓒ Ⓓ
3.		$\frac{3}{x} = \frac{4}{12}$		**3.** Ⓐ Ⓑ Ⓒ Ⓓ
	x		10	
4.	$8^{-3/2}$		$\frac{1}{\sqrt{8^3}}$	**4.** Ⓐ Ⓑ Ⓒ Ⓓ
5.		$2^{x/3} = 8$		**5.** Ⓐ Ⓑ Ⓒ Ⓓ
	x		8	
6.	$4^{1/a}$		$\sqrt[a]{4}$	**6.** Ⓐ Ⓑ Ⓒ Ⓓ
7.	The distance between points (4, 2) and (3, 7)		The distance between points (4, −2) and (3, −7)	**7.** Ⓐ Ⓑ Ⓒ Ⓓ
8.		The ratio of 2 to 3 is equal to the ratio of 6 to $\sqrt{t}$.		**8.** Ⓐ Ⓑ Ⓒ Ⓓ
	t		9	
9.		$\sqrt{(-2-1)^2 + (2-y)^2} = 5$		**9.** Ⓐ Ⓑ Ⓒ Ⓓ
	y		7	
10.	$8\sqrt{3} - \sqrt{27}$		$6\sqrt{3}$	**10.** Ⓐ Ⓑ Ⓒ Ⓓ

Practice for College Entrance Tests **Chapter 8**	**Form A**	Name ______________

Fill in the circle of the letter that corresponds to the correct answer.

1. The figure below has five squares with sides of length a and five squares with sides of length b. What is the length of x in terms of a and b? — **1.** Ⓐ Ⓑ Ⓒ Ⓓ Ⓔ

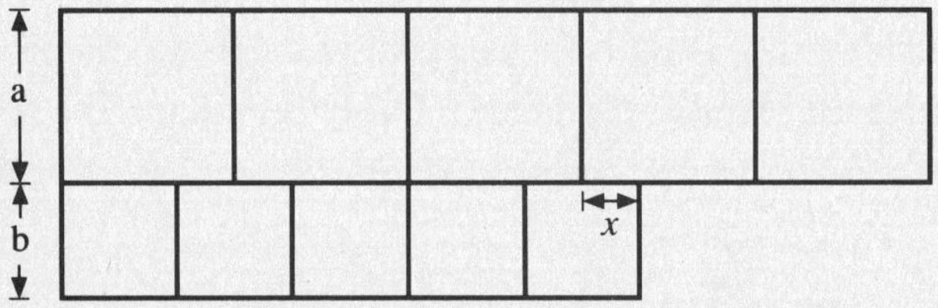

A. $5a - 5b$ **B.** $4a - 5b$ **C.** $5b - 5a$
D. $5b - 4a$ **E.** $5b - 3a$

2. A train leaves the station headed due east averaging 50 miles per hour. A second train headed due west leaves the same station an hour and a half later. If the second train averages 55 miles per hour, when will the two trains be 500 miles apart? — **2.** Ⓐ Ⓑ Ⓒ Ⓓ Ⓔ

A. $\approx$ 4 hours after the first train leaves the station
B. $\approx 5\frac{1}{2}$ hours after the first train leaves the station
C. $\approx$ 7 hours after the first train leaves the station
D. $\approx 5\frac{1}{2}$ hours after the second train leaves the station
E. $\approx$ 7 hours after the second train leaves the station

3. A 2-foot wide sidewalk borders a rectangular garden that has an area of 100 square feet. What is the area, in square feet, covered by the sidewalk? — **3.** Ⓐ Ⓑ Ⓒ Ⓓ Ⓔ

A. 96 **B.** 132 **C.** 196 **D.** 224
E. Cannot be determined by the given information

In 4, use the graph below.

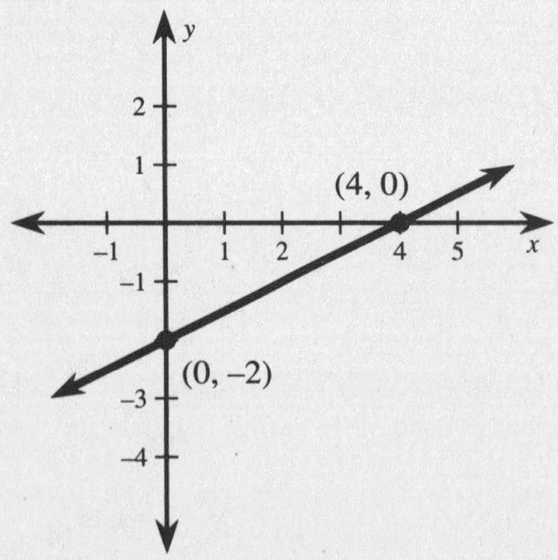

4. What is the slope of the line? — **4.** Ⓐ Ⓑ Ⓒ Ⓓ Ⓔ

A. -2 **B.** $-\frac{1}{2}$ **C.** $\frac{1}{6}$ **D.** $\frac{1}{2}$ **E.** 2

Practice for College Entrance Tests

Chapter 8

Form B

Name ________________

Solve the problem and write your answer at the top of the grid on the separate answer sheet. Then fill in the corresponding ellipses in the answer grid. (Your answer can be an integer, a decimal number, or a fraction.)

1. A box with a square base and a height of 4 inches has a volume of 25 cubic inches. What is the surface area in square inches of the box?

2. If $a = 4$, what is the value of $(3a)^2$?

3. A liter of pure water is added to a 3 liter solution that is 40% alcohol. What is the percent of alcohol after the water is added?

4. A 4-foot tree casts a shadow that is 2 feet long. A neighboring tree casts a shadow of 6 feet. How tall, in feet, is the neighboring tree?

5. A customer purchases a shirt on sale at 20% off and pays a 5% sales tax on the purchase price. If the total cost was \$21.00, what was the original price of the shirt in dollars?

6. If p is 4 more than the quotient of y and 3 and $y = 9$, what is the value of p?

7. If $\frac{x}{y} = 4$, $\frac{y}{z} = 5$, and $z = 3$, what is the value of x?

8. When D dollars is divided equally among 5 people, there is a remainder of \$3. What is the remainder when $(D + 8)$ dollars is divided equally among 5 people?

9. You invest \$5000 in two mutual funds paying 5% annual interest and 8% annual interest. The interest earned for one year was \$340. How much of the interest, in dollars, came from the mutual fund paying 8% annual interest?

10. If $(2.71 \times 10^4)(4.0 \times 10^5) = 1.084 \times 10^n$, then $n =$ [?]

11. If $x + 2y = 10$ and $x - y = -2$ then $y =$ [?]

12. A circle with an area 121π has a diameter of [?]

Practice for College Entrance Tests
Chapter 9

Form A Name ______________________

Fill in the circle of the letter that corresponds to the correct answer.

1. What are the possible values for r in the equation $x^2 - 5x - 36 = (x + r)(x - 9)$?

   A. 12 B. 6 C. 4 D. 4 and -4 E. -4

   1. Ⓐ Ⓑ Ⓒ Ⓓ Ⓔ

2. Which equation has roots of $x = \frac{2}{5}$ and $x = -\frac{4}{5}$?

   A. $(5x + 2)(5x + 4) = 0$ B. $(5x - 2)(5x - 4) = 0$
   C. $(2x - 5)(4x + 5) = 0$ D. $(5x - 2)(5x + 4) = 0$
   E. $(x + \frac{2}{5})(x - \frac{4}{5}) = 0$

   2. Ⓐ Ⓑ Ⓒ Ⓓ Ⓔ

3. Factor completely $3a - 8a^2 + 4a^3$.

   A. $(3a - 2a^2)(1 - 2a)$ B. $a(3 - 4a)(1 - a)$
   C. $(3 - 2a)(a - 2a^2)$ D. $a(3 - 2a)(1 - 2a)$
   E. $a(3 - 8a + 4a^2)$

   3. Ⓐ Ⓑ Ⓒ Ⓓ Ⓔ

4. If $(3x - 2)(x + c) = 3x^2 - 8x - 2c$, what is the value of c?

   A. -10 B. -6 C. -2 D. 2
   E. Cannot be determined from the given information.

   4. Ⓐ Ⓑ Ⓒ Ⓓ Ⓔ

5. For all $x \neq 2$, which of the following is equivalent to $\frac{x^2 + 8x + 12}{x + 2}$?

   A. $x + 2$ B. $x + 4$ C. $x + 6$ D. $x + 8$ E. $x + 12$

   5. Ⓐ Ⓑ Ⓒ Ⓓ Ⓔ

6. What is the sum of all the values of x that satisfy the equation $3x^2 + 3x - 18 = 0$?

   A. -6 B. -5 C. -1 D. 1 E. 5

   6. Ⓐ Ⓑ Ⓒ Ⓓ Ⓔ

7. If $(2x + s)(3x - 2) = 6x^2 + 11x - 2s$, what is the value of s?

   A. $\frac{7}{3}$ B. $\frac{11}{3}$ C. 5 D. $\frac{17}{2}$
   E. Not enough given information

   7. Ⓐ Ⓑ Ⓒ Ⓓ Ⓔ

8. Which of the following are solutions to $6x^2 + x - 12 = 0$?

   A. $\{-\frac{3}{2}, \frac{4}{3}\}$ B. $\{-\frac{2}{3}, \frac{1}{2}\}$ C. $\{\frac{2}{3}, \frac{1}{2}\}$
   D. $\{\frac{3}{2}, -\frac{4}{3}\}$ E. $\{\frac{3}{2}, \frac{4}{3}\}$

   8. Ⓐ Ⓑ Ⓒ Ⓓ Ⓔ

9. Which of the following quadratic equations has $-\frac{1}{2}$ and $-\frac{2}{3}$ as solutions?

   A. $3x^2 - x + 1 = 0$ B. $3x^2 + x + 1 = 0$
   C. $6x^2 - 7x - 2 = 0$ D. $6x^2 - 7x + 2 = 0$
   E. $6x^2 + 7x + 2 = 0$

   9. Ⓐ Ⓑ Ⓒ Ⓓ Ⓔ

Practice for College Entrance Tests

Chapter 9

Form B **Name** ______________________

If Column A is greater than Column B, fill in the circle for A.
If Column B is greater than Column A, fill in the circle for B.
If Column A is equal to Column B, fill in the circle for C.
If there is not enough information, fill in the circle for D.
Information centered between the columns concerns the quantities in both columns.

	Column A		Column B	
1.		$x^2 - 7x + 12 = 0$		**1.** Ⓐ Ⓑ Ⓒ Ⓓ
	5		x	
2.		$x^2 - 36 = 0$		**2.** Ⓐ Ⓑ Ⓒ Ⓓ
	x		-6	
3.		$x^2 + nx + 9 = 0$ $n > 0$		**3.** Ⓐ Ⓑ Ⓒ Ⓓ
	n		6	
4.		$x \neq -2$		**4.** Ⓐ Ⓑ Ⓒ Ⓓ
	$\frac{x^2 + 5x + 6}{x + 2}$		$x + 2$	
5.		$x \neq 1$		**5.** Ⓐ Ⓑ Ⓒ Ⓓ
	3		$\frac{3x^2 - 6x + 3}{(x - 1)^2}$	
6.		$(s + 3)(s + r) = s^2 + 8s + 15$		**6.** Ⓐ Ⓑ Ⓒ Ⓓ
	r		7	
7.		t is a positive integer and $t(t + 1)(t + 2) = 60$		**7.** Ⓐ Ⓑ Ⓒ Ⓓ
	$t + 2$		6	
8.	$p + q$		$(2p + q) - (p - q)$	**8.** Ⓐ Ⓑ Ⓒ Ⓓ
9.		$c < 0$		**9.** Ⓐ Ⓑ Ⓒ Ⓓ
	0		$c^4 - c^3 + c^2$	
10.	$y^2 + 3y - 4$		$4 - 3y - y^2$	**10.** Ⓐ Ⓑ Ⓒ Ⓓ

Practice for College Entrance Tests

Chapter 10

Form A

Name ______________________

Fill in the circle of the letter that corresponds to the correct answer.

1. If $\frac{3+4}{x} = \frac{3+x}{4}$, which of the following could be a value of x?

   A. -7 B. -4 C. 1   D. 7   E. 28

   1. Ⓐ Ⓑ Ⓒ Ⓓ Ⓔ

2. If $b = 3$, then $\frac{b^3 - b}{2b} = \boxed{?}$

   A. 24   B. 8   C. 6   D. 5   E. 4

   2. Ⓐ Ⓑ Ⓒ Ⓓ Ⓔ

3. If $x^* = \frac{x+3}{x+1}$ and x is positive, then $2\left(\frac{1}{x^*}\right) = \boxed{?}$

   A. $\frac{x+1}{x+3}$ B. $\frac{2x+1}{x+3}$ C. $\frac{2x+2}{x+3}$
   D. $\frac{2x+3}{x+1}$ E. $\frac{2x+6}{x+1}$

   3. Ⓐ Ⓑ Ⓒ Ⓓ Ⓔ

4. For $v^* = \frac{v^2+2}{v-3}$ and $v \neq 3$, what does 2^* equal?

   A. -6 B. $-\frac{4}{3}$ C. $\frac{6}{5}$ D. $\frac{4}{3}$ E. 6

   4. Ⓐ Ⓑ Ⓒ Ⓓ Ⓔ

5. For which values of x is $\frac{3x}{x+1} + \frac{x^2+1}{x-3} = 0$ undefined?

   A. -3 and 1   B. -1 and 3   C. 0   D. 1   E. -3

   5. Ⓐ Ⓑ Ⓒ Ⓓ Ⓔ

6. If $\frac{2}{x+3} = \frac{1}{3}$, then $x = \boxed{?}$

   A. 0   B. 1   C. 2   D. 3   E. 4

   6. Ⓐ Ⓑ Ⓒ Ⓓ Ⓔ

7. Malcolm can mow and trim the yard in 45 minutes. C. J. can do the job in 30 minutes. How many minutes would it take both of them working together?

   A. 15   B. 18   C. 22.5   D. 37.5   E. 75

   7. Ⓐ Ⓑ Ⓒ Ⓓ Ⓔ

8. If $\frac{x+7}{3} = \frac{1-5}{x}$, which of the following could be the value(s) for x?

   I. -3 II. -4 III. 4   IV. 2

   A. I. only   B. III. only   C. IV. only
   D. I. and II.   E. I. and III.

   8. Ⓐ Ⓑ Ⓒ Ⓓ Ⓔ

Practice for College Entrance Tests

Chapter 10

Form B Name ____________________

Solve the problem and write your answer at the top of the grid on the separate answer sheet. Then fill in the corresponding ellipses in the answer grid. (Your answer can be an integer, a decimal number, or a fraction.)

1. For which value of a is the equation $\frac{3a}{a-1} = 6$ undefined?

2. If $\frac{x-3}{x} = \frac{x-2}{x+2}$ and $x \neq -2$ or 0, then $x = \boxed{?}$

3. A rectangular patio made of bricks has sides of length 16 ft and 25 ft. If the bricks are rearranged to make a square patio, what is the length, in feet, of each side?

4. What is the sum of all the solutions to $\frac{x-1}{x+15} = \frac{1}{x}$?

5. Using a calculator, Jamie tried to divide a number, N, by 8. Instead, the button "9" was pushed and the result was 2 less than what it should have been. What is the value of N?

6. Grid in the solution to $\frac{3x-4}{6} = 2x - 1$.

7. It takes Janine 30 minutes to wash dishes and clean up the kitchen after supper. Her younger brother takes 40 minutes. How long, in minutes, would it take for Janine and her brother, working together, to wash dishes and clean up the kitchen? (Round your answer to one decimal place.)

8. Grid in one of the values for which the equation $\frac{3x}{x-2} = \frac{x+2}{x-1}$ is undefined.

9. If $\boxed{w} = \frac{w^2 - 3}{w+5}$ and $w \neq -5$, then what does $\boxed{2}$ equal?

10. The variables t and h vary inversely. When $t = 6$, $h = 36$. What is the value of t when $h = 27$?

11. What is the sum of the solutions to the equation $\frac{x}{2x-9} = \frac{2}{x-5}$?

Practice for College Entrance Tests
Chapter 11

Form A Name ____________________

Fill in the circle of the letter that corresponds to the correct answer.

1. The graph of $y = x^2 + 2$ is a parabola with $x = 0$ as its axis of symmetry. Which of the following are the (x, y) coordinates of the point on the parabola that is symmetric to the point with coordinates (2, 6)? 1. Ⓐ Ⓑ Ⓒ Ⓓ Ⓔ

   A. (3, 11) B. (1, 3) C. (0, 2)
   D. (−1, 3) E. (−2, 6)

2. What is the equation of a circle with the center at the origin and a radius of 4 units? 2. Ⓐ Ⓑ Ⓒ Ⓓ Ⓔ

   A. $x + y = 4$ B. $x^2 + y^2 = 4$ C. $x^2 + y^2 = 16$
   D. $\frac{x^2}{4} + \frac{y^2}{4} = 1$ E. $x + y = 8$

3. If the equation $y - x^2 + 2 = 0$ were graphed in the standard coordinate plane, the graph would be which of the following? 3. Ⓐ Ⓑ Ⓒ Ⓓ Ⓔ

   A. Circle B. Ellipse C. Parabola
   D. Hyperbola E. Straight line

4. In the coordinate plane, which of the following is the equation of an ellipse with its center at the origin, vertices of (4, 0) and (−4, 0), and covertices of (0, 3) and (0, −3)? 4. Ⓐ Ⓑ Ⓒ Ⓓ Ⓔ

   A. $\frac{x^2}{3} + \frac{y^2}{4} = 1$ B. $\frac{x^2}{4} + \frac{y^2}{3} = 1$
   C. $\frac{x^2}{8} + \frac{y^2}{6} = 1$ D. $\frac{x^2}{9} + \frac{y^2}{16} = 1$
   E. $\frac{x^2}{16} + \frac{y^2}{9} = 1$

5. What is the equation of a circle with the center at (−1, 3) and a radius of 6 units? 5. Ⓐ Ⓑ Ⓒ Ⓓ Ⓔ

   A. $x + y = 4$ B. $\frac{x^2}{1} + \frac{y^2}{9} = 36$
   C. $x^2 + y^2 = 36$ D. $(x - 1)^2 + (y + 3)^2 = 36$
   E. $(x + 1)^2 + (y - 3)^2 = 36$

6. If the equation $3x^2 + 2y^2 = 6$ were graphed in the standard coordinate plane, the graph would be which of the following? 6. Ⓐ Ⓑ Ⓒ Ⓓ Ⓔ

   A. Circle B. Ellipse C. Parabola
   D. Hyperbola E. Straight line

Practice for College Entrance Tests

Chapter 11

Form B Name ______________

If Column A is greater than Column B, fill in the circle for A.
If Column B is greater than Column A, fill in the circle for B.
If Column A is equal to Column B, fill in the circle for C.
If there is not enough information, fill in the circle for D.
Information centered between the columns concerns the quantities in both columns.

	Column A		Column B	
1.		x is an even integer. y is an odd integer.		**1.** Ⓐ Ⓑ Ⓒ Ⓓ
	$2x$		$2y$	
2.	$\frac{1}{3-\frac{9}{5}}$		$3+\frac{9}{5}$	**2.** Ⓐ Ⓑ Ⓒ Ⓓ
3.		$4 \times 10^n = (8 \times 10^4)(5 \times 10^{-1})$		**3.** Ⓐ Ⓑ Ⓒ Ⓓ
	4		n	
4.		 $b = 6,\ \frac{a}{b} = \frac{2}{3}$		**4.** Ⓐ Ⓑ Ⓒ Ⓓ
	a		3	
5.	The time it would take a cyclist to travel 15 miles at an average speed of 10 miles per hour		The time it would take a cyclist to travel 10 miles at an average speed of 15 miles per hour	**5.** Ⓐ Ⓑ Ⓒ Ⓓ
6.		$(y-2)(y+c) = y^2 - 5y + 6$		**6.** Ⓐ Ⓑ Ⓒ Ⓓ
	-5		c	
7.	$\frac{1010}{50}$		$\frac{101}{5}$	**7.** Ⓐ Ⓑ Ⓒ Ⓓ
8.		Let the operation $\circ$ have the property $p \circ q = 2p - q$.		**8.** Ⓐ Ⓑ Ⓒ Ⓓ
	$\frac{3}{4} \circ \frac{1}{4}$		$\frac{7}{4}$	
9.		4 more than $3b$ is 89.		**9.** Ⓐ Ⓑ Ⓒ Ⓓ
	30		b	

Practice for College Entrance Tests

Chapter 12

Form A **Name** ______________________

Fill in the circle of the letter that corresponds to the correct answer.

1. What is the decimal representation of $8 + \frac{7}{100} + \frac{7}{10{,}000}$?   1. Ⓐ Ⓑ Ⓒ Ⓓ Ⓔ

A. 8.77 **B.** 8.707 **C.** 8.077 **D.** 8.0707 **E.** 8.0077

2. An oatmeal cookie recipe calls for 1 cup sugar, 2 cups oatmeal, and $\frac{3}{4}$ cup nuts. What is the ratio of sugar to oatmeal to nuts?   2. Ⓐ Ⓑ Ⓒ Ⓓ Ⓔ

A. 1:2:3 **B.** 1:8:3 **C.** 3:6:4 **D.** 4:2:3 **E.** 4:8:3

3. Which of the following systems of equations represents the graph?   3. Ⓐ Ⓑ Ⓒ Ⓓ Ⓔ

A. $\begin{cases} -6x + 4y \le 24 \\ 4x + 3y \ge 2 \\ y \le -2 \end{cases}$ **B.** $\begin{cases} -6x + 4y \ge 24 \\ 4x + 3y \ge 2 \\ y \ge -2 \end{cases}$

C. $\begin{cases} -6x + 4y \ge 24 \\ 4x + 3y \le 2 \\ y \le -2 \end{cases}$ **D.** $\begin{cases} -6x + 4y \le 24 \\ 4x + 3y \le 2 \\ y \le -2 \end{cases}$

E. $\begin{cases} -6x + 4y \le 24 \\ 4x + 3y \le 2 \\ y \ge -2 \end{cases}$

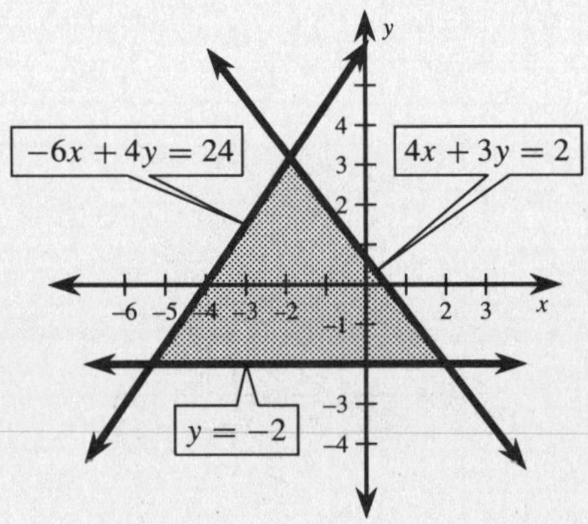

4. If $x + y = 8p$ and $x - y = 6q$, then x is [?]   4. Ⓐ Ⓑ Ⓒ Ⓓ Ⓔ

A. $7pq$ **B.** $4p + 3q$ **C.** pq
D. $4p - 3q$ **E.** $8p + 6q$

5. A person is hired for a job that pays \$500 per month plus a 10% raise after each month. In the fourth month, how much will the person earn?   5. Ⓐ Ⓑ Ⓒ Ⓓ Ⓔ

A. \$550 **B.** \$600.50 **C.** \$650.50 **D.** \$665.50 **E.** \$700

6. A buffet dinner costs \$6.95 for adults and \$3.95 for children. If a family of 8 paid \$46.60, how many paid the adult price?   6. Ⓐ Ⓑ Ⓒ Ⓓ Ⓔ

A. 6 **B.** 5 **C.** 4 **D.** 3 **E.** 2

7. A day care center plans to fence in an outdoor play area with 160 feet of fencing. If the length is 10 feet wider than its width, what is the area in square feet of the play area?   7. Ⓐ Ⓑ Ⓒ Ⓓ Ⓔ

A. 6375 **B.** 1600 **C.** 1575 **D.** 1400 **E.** 1225

Practice for College Entrance Tests Chapter 12

Form B Name ______________________

Solve the problem and write your answer at the top of the grid on the separate answer sheet. Then fill in the corresponding ellipses in the answer grid. (Your answer can be an integer, a decimal number, or a fraction.)

1. The length of a room is 6 feet less than 3 times its width. The area of the room is 360 square feet. Grid in either the length or the width of the room in feet.

2. If $x^2 + 3x - 3 = (x + a)^2 - \frac{21}{4}$, what is the value of a?

In 3 and 4, use the graph and the information below.

M is the midpoint of the line segment $\overline{AB}$ in the coordinate plane.

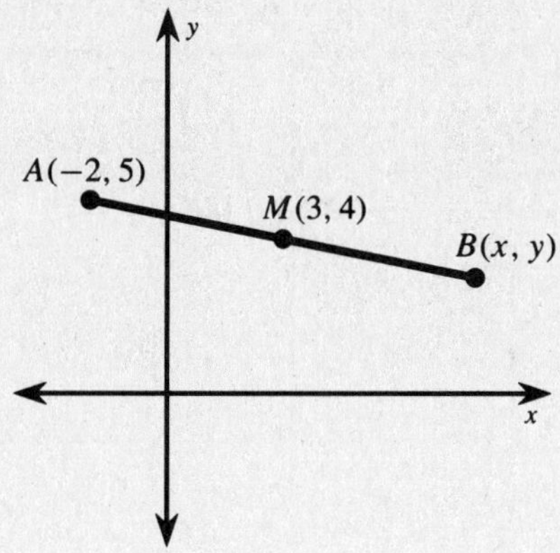

3. What is the x-coordinate of point B?

4. What is the y-coordinate of point B?

5. The store sells a 2-pound bag of walnuts for \$3.96. What is the price per pound, in dollars, of walnuts?

6. The average of 8 numbers is 22. The sum of 5 of these numbers is 80. What is the average of the remaining 3 numbers?

7. Grid in one possible solution of $|x - 10| = 3$.

8. What is the slope of the line $-10x + 6y = 3$?

Practice for College Entrance Tests

Chapter 13

Form A Name ____________________

Fill in the circle of the letter that corresponds to the correct answer.

In 1–5, use the triangle below.

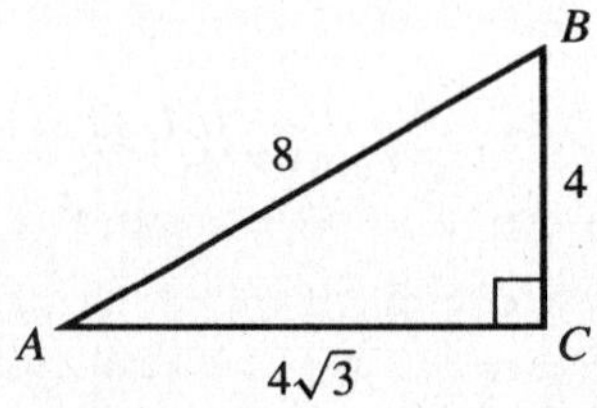

1. The $\sin A =$ [?] — **1.** Ⓐ Ⓑ Ⓒ Ⓓ Ⓔ

A. $\frac{1}{2}$ **B.** $\frac{\sqrt{3}}{3}$ **C.** $\frac{\sqrt{3}}{2}$ **D.** $\frac{2\sqrt{3}}{3}$ **E.** $\sqrt{3}$

2. The $\cos B =$ [?] — **2.** Ⓐ Ⓑ Ⓒ Ⓓ Ⓔ

A. $\frac{1}{2}$ **B.** $\frac{\sqrt{3}}{3}$ **C.** $\frac{\sqrt{3}}{2}$ **D.** $\frac{2\sqrt{3}}{3}$ **E.** $\sqrt{3}$

3. The $\sin B =$ [?] — **3.** Ⓐ Ⓑ Ⓒ Ⓓ Ⓔ

A. $\frac{1}{2}$ **B.** $\frac{\sqrt{3}}{3}$ **C.** $\frac{\sqrt{3}}{2}$ **D.** $\frac{2\sqrt{3}}{3}$ **E.** $\sqrt{3}$

4. The $\tan A =$ [?] — **4.** Ⓐ Ⓑ Ⓒ Ⓓ Ⓔ

A. $\frac{1}{2}$ **B.** $\frac{\sqrt{3}}{3}$ **C.** $\frac{\sqrt{3}}{2}$ **D.** $\frac{2\sqrt{3}}{3}$ **E.** $\sqrt{3}$

5. The $\tan B =$ [?] — **5.** Ⓐ Ⓑ Ⓒ Ⓓ Ⓔ

A. $\frac{1}{2}$ **B.** $\frac{\sqrt{3}}{3}$ **C.** $\frac{\sqrt{3}}{2}$ **D.** $\frac{2\sqrt{3}}{3}$ **E.** $\sqrt{3}$

6. In a right triangle the right angle is labeled C and the other two angles are labeled A and B. Which of the following statements is a true statement about the relationship of $\angle A$ and $\angle B$? — **6.** Ⓐ Ⓑ Ⓒ Ⓓ Ⓔ

A. One and only one could be an obtuse angle
B. One and only one could be an acute angle
C. One could be a right angle
D. Both are obtuse angles
E. Both are acute angles

7. If $0° < x° < 90°$ and $\sin x = \frac{\sqrt{3}}{2}$, then $\cos x =$ [?] — **7.** Ⓐ Ⓑ Ⓒ Ⓓ Ⓔ

A. $\frac{1}{2}$ **B.** $\frac{\sqrt{3}}{3}$ **C.** $\frac{2\sqrt{3}}{3}$ **D.** $\sqrt{3}$ **E.** 2

Practice for College Entrance Tests

Chapter 13

Form B

Name ______________________

If Column A is greater than Column B, fill in the circle for A.
If Column B is greater than Column A, fill in the circle for B.
If Column A is equal to Column B, fill in the circle for C.
If there is not enough information, fill in the circle for D.
Information centered between the columns concerns the quantities in both columns.

	Column A	Column B	
1.	$3x + y$	$x + 3y$	**1.** Ⓐ Ⓑ Ⓒ Ⓓ
2.	The average of 5, 9, and 13	$\frac{27}{2}$	**2.** Ⓐ Ⓑ Ⓒ Ⓓ

3. Alberto's monthly pay is $400 plus 10% of his sales. **3.** Ⓐ Ⓑ Ⓒ Ⓓ

Column A	Column B
Alberto's monthly pay with sales of $5000	$1000

4. A new car valued at $15,000 depreciates $2000 a year for the next 5 years. **4.** Ⓐ Ⓑ Ⓒ Ⓓ

Column A	Column B
$10,000	The value of the car after 5 years

In 5 and 6, use the triangle below.

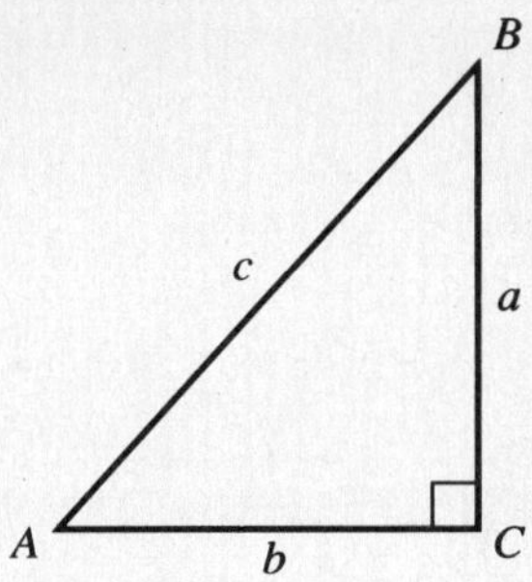

5. $a > b$ **5.** Ⓐ Ⓑ Ⓒ Ⓓ

Column A	Column B
tan A	tan B

	Column A	Column B	
6.	tan A	1	**6.** Ⓐ Ⓑ Ⓒ Ⓓ

Practice for College Entrance Tests

Chapter 14

Form A

Name ______________________

Fill in the circle of the letter that corresponds to the correct answer.

1. For $-\frac{\pi}{2} < \theta < \frac{\pi}{2}$, $\frac{\sin^2\theta + \cos^2\theta}{\tan\theta} = \boxed{?}$ 1. Ⓐ Ⓑ Ⓒ Ⓓ Ⓔ

A. $\sin\theta$ B. $\cos\theta$ C. $\tan\theta$
D. $\cot\theta$ E. $\csc\theta$

2. What is the value of $\left(\cos\frac{\pi}{4}\right)\left(\cos\frac{\pi}{4}\right) + \left(\sin\frac{\pi}{4}\right)\left(\sin\frac{\pi}{4}\right)$? 2. Ⓐ Ⓑ Ⓒ Ⓓ Ⓔ

A. $\frac{2}{\sqrt{2}}$ B. 1 C. $\frac{1}{\sqrt{2}}$ D. $\frac{1}{4}$ E. 0

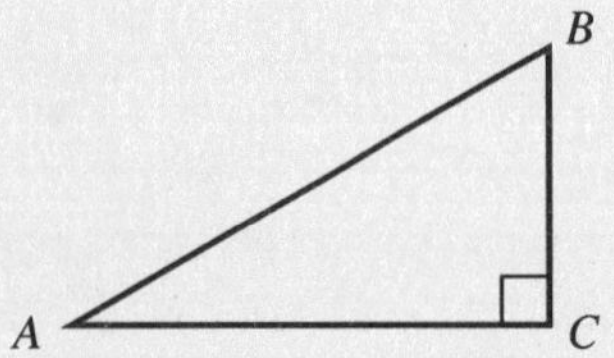

3. If $\angle A$ is 30°, what is the ratio of $\overline{AC}$ to $\overline{BC}$? 3. Ⓐ Ⓑ Ⓒ Ⓓ Ⓔ

A. $\frac{1}{2}$ B. $\frac{\sqrt{3}}{3}$ C. $\frac{\sqrt{3}}{2}$ D. $\frac{2\sqrt{3}}{3}$ E. $\sqrt{3}$

4. For $-\frac{\pi}{2} < \theta < \frac{\pi}{2}$, $\frac{1 + \tan^2\theta}{\sec\theta} = \boxed{?}$ 4. Ⓐ Ⓑ Ⓒ Ⓓ Ⓔ

A. $\sin\theta$ B. $\cos\theta$ C. $\tan\theta$
D. $\csc\theta$ E. $\sec\theta$

In 5, use the graph and the information below.

One cycle of the cosine curve is shown to the right. The period is 8 seconds, after which the values repeat. The curve has a height of *r* at 0 and 8 seconds, *s* at 1 and 7 seconds, *t* at 2 and 6 seconds, *u* at 3 and 5 seconds, and *v* at 4 seconds.

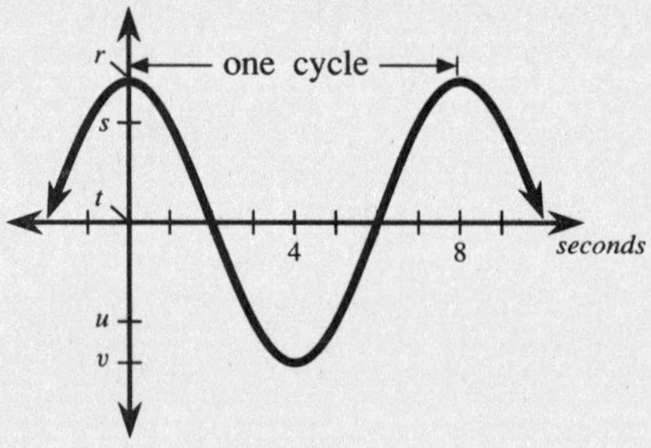

5. What is the height of the curve after 52 seconds? 5. Ⓐ Ⓑ Ⓒ Ⓓ Ⓔ

A. r B. s C. t D. u E. v

Practice for
College Entrance Tests
Chapter 14

Form B

Name ______________________

Solve the problem and write your answer at the top of the grid on the separate answer sheet. Then fill in the corresponding ellipses in the answer grid. (Your answer can be an integer, a decimal number, or a fraction.)

1. Mr. Hartwell's gross monthly income is \$2000. Deductions include 15% federal tax, 2.5% state tax, and 1% local tax. What is Mr. Hartwell's net monthly income, in dollars, after these deductions?

2. A motorboat travels 30 miles in 40 minutes. What is its speed in miles per hour?

3. The sides of a triangle are 6, 9, and 14 units long. The longest side of a similar triangle is 35 units. What is the perimeter of the second triangle?

4. The formula for calculating simple interest is $I = Pr$ where I is the interest earned, P is the initial amount invested, and r is the fixed interest rate. If \$600 was invested and \$45 was earned in interest, what was the interest rate, expressed as a percentage?

5. Admission to the annual teacher-student volleyball game was \$2 for adults and \$1 for students. A total of \$500 was collected. If 400 people attended, how many were students?

6. A line with a slope of -2 goes through the point $(3, 4)$. If the point $(-1,\ b)$ is also on the line, what is b?

7. If $\left(\frac{1}{2} - \frac{2}{3}\right) + \left(\frac{1}{4} + \frac{1}{8}\right)$ is calculated and reduced to its simplest terms, what is the denominator of the resulting fraction?

8. If 7 is 3 more than $\frac{1}{2}$ of x, what is the value of x?

9. If $\sqrt[a]{72} = 2\sqrt[a]{9}$, what is the value of a?

10. If $(x + 3)(x - c) = x^2 + x - 6$, then what is the value of c?

11. A line segment has end points at $(-4.5,\ 3)$ and $(7,\ -2)$. Grid in the value of the x-coordinate of the midpoint of the line segment.

Practice for College Entrance Tests
Chapter 15
Form A **Name** ____________

Fill in the circle of the letter that corresponds to the correct answer.

1. Slips of paper numbered 1 to 15 are placed in a hat. What is the probability that the first number drawn is a multiple of 2? **1.** Ⓐ Ⓑ Ⓒ Ⓓ Ⓔ

A. $\frac{1}{15}$ **B.** $\frac{2}{15}$ **C.** $\frac{7}{15}$ **D.** $2\frac{1}{2}$ **E.** $\frac{8}{15}$

2. How many different ways can the letters o, p, s, and t be arranged? **2.** Ⓐ Ⓑ Ⓒ Ⓓ Ⓔ

A. 4 **B.** 16 **C.** 24 **D.** 64 **E.** 256

3. A vendor at a carnival has 7 red balloons, 6 blue balloons, 8 yellow balloons, 4 green balloons, and 5 orange balloons. If the vendor randomly picks one balloon to sell, what is the probability that the balloon is blue or green? **3.** Ⓐ Ⓑ Ⓒ Ⓓ Ⓔ

A. $\frac{2}{15}$ **B.** $\frac{1}{5}$ **C.** $\frac{1}{4}$ **D.** $\frac{1}{3}$ **E.** $\frac{1}{2}$

4. You dial your friend's phone number but can't remember the last two digits. What is the probability that you dial the right number? **4.** Ⓐ Ⓑ Ⓒ Ⓓ Ⓔ

A. $\frac{1}{100}$ **B.** $\frac{1}{50}$ **C.** $\frac{1}{20}$ **D.** $\frac{1}{10}$ **E.** $\frac{1}{5}$

In 5 and 6, use the following information.

You are taking a multiple-choice test where each question may be answered *a, b, c, d,* or *e*.

5. You know the answer is not **b**. What is the probability that you will choose the right answer? **5.** Ⓐ Ⓑ Ⓒ Ⓓ Ⓔ

A. $\frac{1}{5}$ **B.** $\frac{1}{4}$ **C.** $\frac{1}{3}$ **D.** $\frac{1}{2}$ **E.** $\frac{2}{3}$

6. You know the answer is not **b** or **c**. What is the probability that you will choose the right answer? **6.** Ⓐ Ⓑ Ⓒ Ⓓ Ⓔ

A. $\frac{1}{5}$ **B.** $\frac{1}{4}$ **C.** $\frac{1}{3}$ **D.** $\frac{1}{2}$ **E.** $\frac{2}{3}$

Practice for College Entrance Tests
Chapter 15

Form B Name ______________________

If Column A is greater than Column B, fill in the circle for A.
If Column B is greater than Column A, fill in the circle for B.
If Column A is equal to Column B, fill in the circle for C.
If there is not enough information, fill in the circle for D.
Information centered between the columns concerns the quantities in both columns.

	Column A	Column B	
1.	A jar contains 12 green marbles, 6 yellow marbles, and 7 orange marbles.		**1.** Ⓐ Ⓑ Ⓒ Ⓓ
	The probability that a randomly picked marble is green	The probability that a randomly picked marble is not green	

In 2 and 3, use the following information.

One card is drawn from a deck of cards numbered 1 to 30.

	Column A	Column B	
2.	The probability that the card drawn has a number greater than 15	The probability that the card drawn has a number less than 16	**2.** Ⓐ Ⓑ Ⓒ Ⓓ
3.	The probability that the card drawn has a one-digit number	The probability that the the card drawn has a two-digit number	**3.** Ⓐ Ⓑ Ⓒ Ⓓ
4.	A fair coin is tossed 3 times.		**4.** Ⓐ Ⓑ Ⓒ Ⓓ
	The chances of getting 3 heads	The chances of getting 3 tails	
5.	One bag of marbles contains 6 white marbles and 7 black marbles. A second bag contains 8 black marbles and 9 white marbles.		**5.** Ⓐ Ⓑ Ⓒ Ⓓ
	The probability of picking a white marble from the first bag	The probability of of picking a black marble from the second bag	
6.	$ab = 36.$		**6.** Ⓐ Ⓑ Ⓒ Ⓓ
	a	b	

Answers to Tests

■ Mid-Chapter Test 1-A

1. $-\frac{1}{2}$

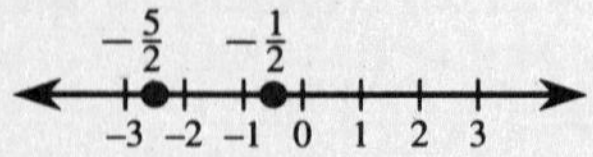

2. $7, \frac{2}{3}, \frac{3}{5}, 0, -\sqrt{3}, -\frac{7}{2}$

3. 13 **4.** $-\frac{4}{3}$ **5.** $60 - 9\pi$ or ≈ 31.7

6. 12.70 **7.** $x = \frac{9}{4}$ **8.** $x = -\frac{1}{5}$

■ Mid-Chapter Test 1-B

1. -1

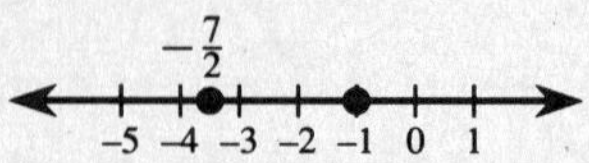

2. $\frac{5}{3}, \frac{3}{2}, \frac{2}{3}, -\frac{5}{3}, -\frac{7}{4}, -5$

3. 10 **4.** $\frac{5}{4}$ **5.** $96 - 16\pi$ or ≈ 45.7

6. 15.47 **7.** $x = -\frac{14}{3}$ **8.** $x = \frac{3}{7}$

■ Chapter Test 1-A

1. 5

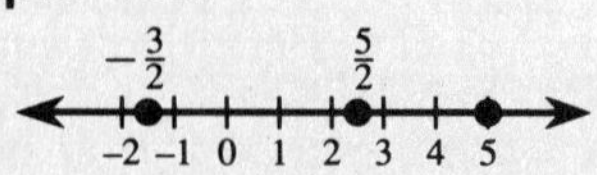

2. $-\frac{7}{2}, -2, -1, 0, \frac{3}{5}, \frac{4}{3}, \frac{5}{3}$

3. Associative property for multiplication

4. Inverse property of addition

5. 27 **6.** 68 **7.** 44 **8.** 21.59

9. $x = -\frac{5}{8}$ **10.** $x = \frac{4}{7}$ **11.** 15

12. $x^3 3^n$ **13.** 0 **14.** \$39,000

15. $F = \frac{9}{5}C + 32$ **16.** $q = \frac{14}{p^2 - 3}$

17. $x < -1$

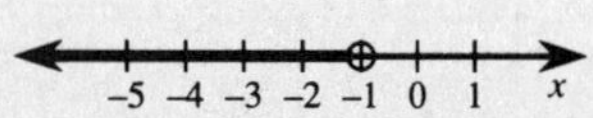

18. No

19. $-1 \leq x \leq \frac{7}{3}$

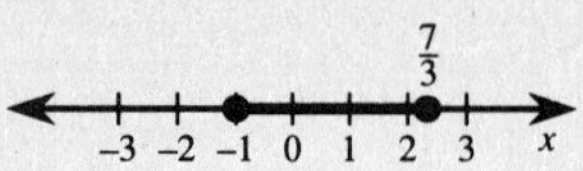

20. $0 \leq A + B \leq 15,000$

■ Chapter Test 1-B

1. 3

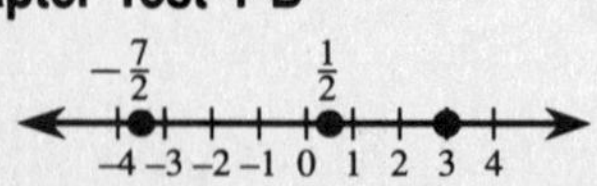

2. $\frac{5}{3}, \frac{4}{3}, \frac{3}{5}, 0, -1, -2, -\frac{7}{2}$

3. Commutative property of addition

4. Associative property of addition

5. -22 **6.** 50 **7.** 96 **8.** 22.62

9. $x = -\frac{3}{10}$ **10.** $x = \frac{5}{2}$ **11.** 20

12. $3^3 x^n$ **13.** 0 **14.** \$37,000

15. $C = \frac{5}{9}(F - 32)$ **16.** $s = \frac{7}{r^2 - 5}$

17. $x \geq 2$

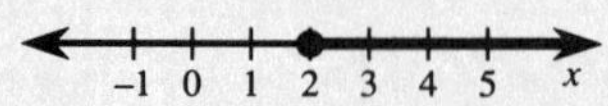

18. No

19. $-1 \leq x \leq 4$

20. $0 \leq A + B \leq 25,000$

■ Chapter Test 1-C

1. $\frac{5}{3}$

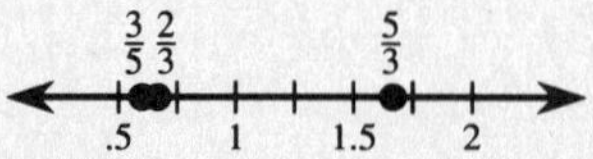

2. $-1, -\frac{2}{3}, -\frac{1}{3}, 0, \frac{3}{5}, \frac{2}{3}$

3. Inverse property of multiplication

4. Commutative property of addition

5. 8 **6.** 92 **7.** $\frac{5}{2}$ **8.** 5.57

9. $x = -\frac{9}{4}$ **10.** $x = \frac{20}{9}$ **11.** 20

12. $(2x)^n$ **13.** -270 **14.** 1,081,200

15. $P = \frac{A}{1+rt}$ **16.** $z = \frac{y}{y^2+7}$

17. $x < 1$

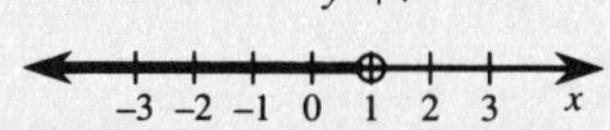

18. Yes

19. $x < 2$ or $x > 3$

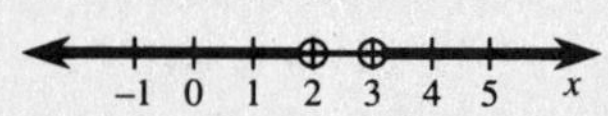

20. $3,000 \leq A \leq 17,000$

■ Mid-Chapter Test 2-A

1. Quad. II, $(-2, 1)$
Quad. I, $(3, 2)$
Quad. III, $(-1, -3)$

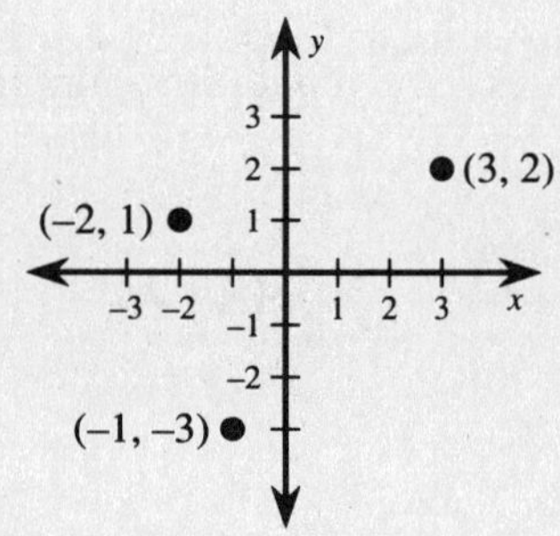

2. 13, 10, 7, 4, 1, -2

3. Slope: 2
y-intercept: $(0, -1)$

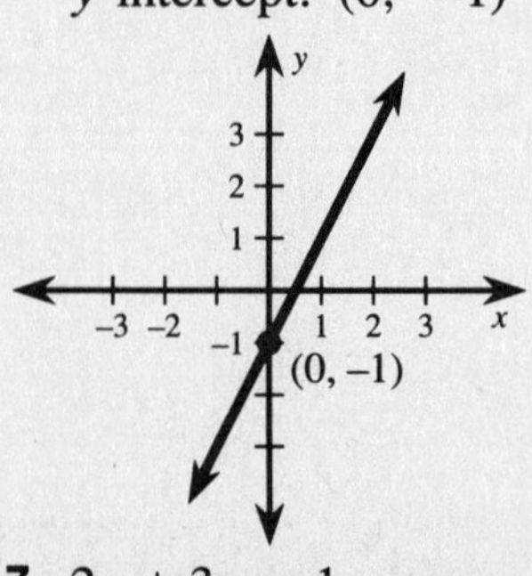

4. $y = -2$

5. $2x + 4y = 7$
$4x + 8y = 5$

6. $y = -\frac{5}{7}x - \frac{4}{7}$

7. $2x + 3y = 1$

■ Mid-Chapter Test 2-B

1. Quad. I, (3, 3)
   Quad. IV, (2, −2)
   Quad. III, (−1, −3)

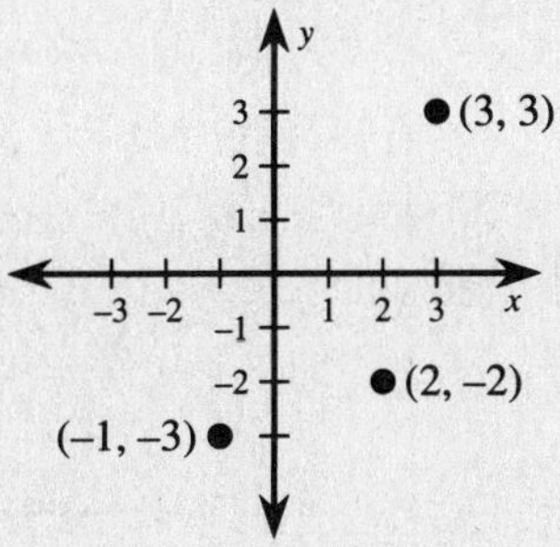

2. −16, −12, −8, −4, 0, 4
3. Slope: −3
   y-intercept: (0, 3)

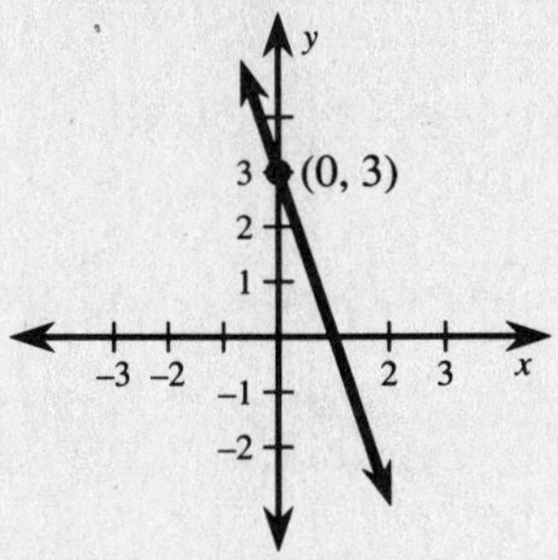

4. $x = 5$
5. $2x + 4y = 7$
   $4x - 2y = 7$
6. $y = \frac{6}{7}x - \frac{4}{7}$
7. $x - 2y = -7$

■ Chapter Test 2-A

1. Quadrant III 2. $x = -5$
3.

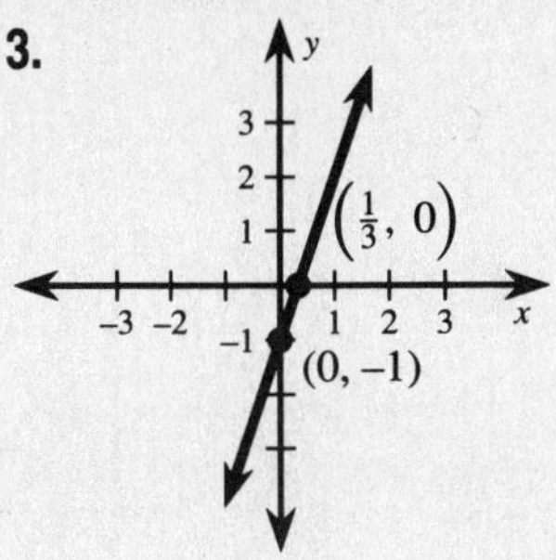

4. $\frac{5}{3}$
5. $y = \frac{5}{3}x - 2$
6. Parallel
7. (4, 0), (0, −5)
8. $y = 5x - \frac{7}{3}$; Slope: 5; y-intercept: $(0, -\frac{7}{3})$
9.

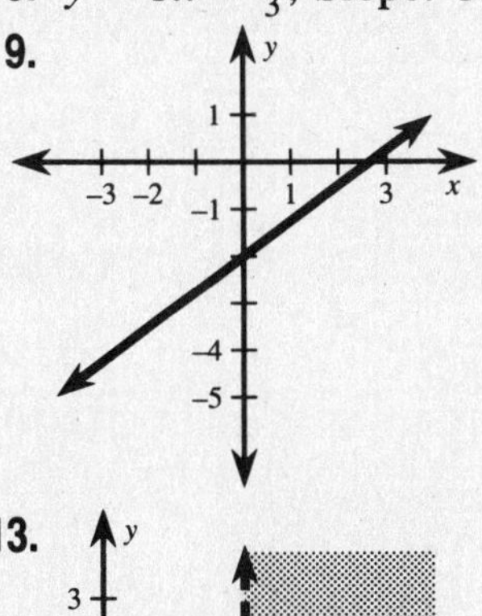

10. $y = -24x$
11. $C = 0.20x + 25$
12. Yes
13.

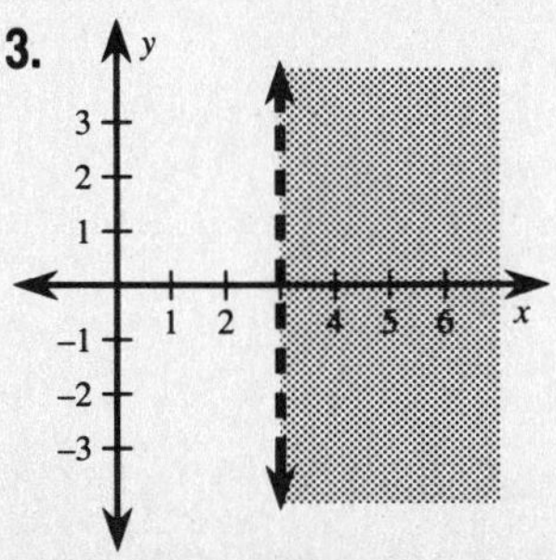

14.

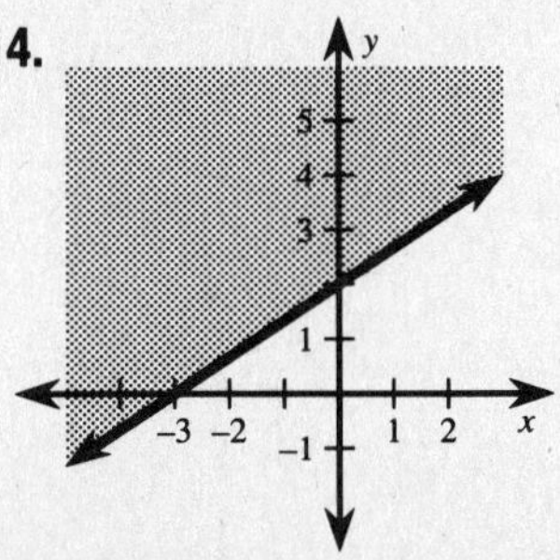

15. (3, −3) 16.

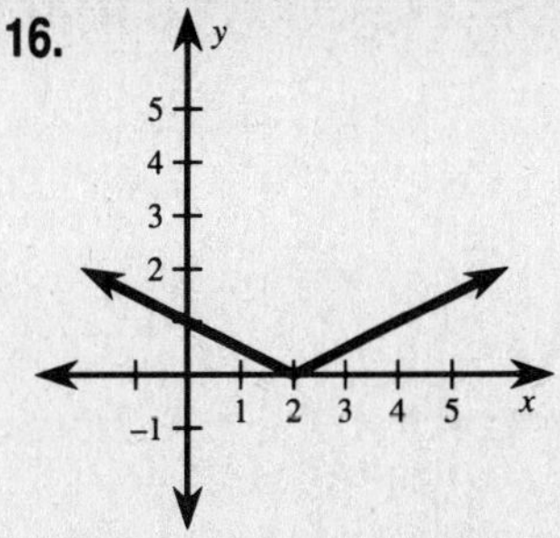

17. 1994, 1998 18. Negative correlation

■ Chapter Test 2-B

1. Quadrant IV 2. $y = -3$
3.

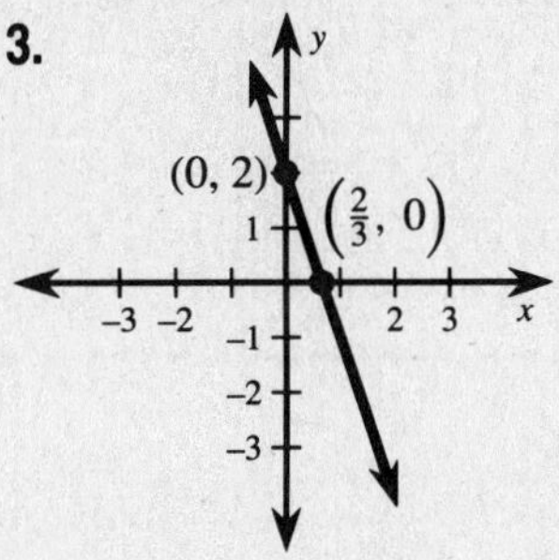

4. −1
5. $y = \frac{2}{3}x + 3$
6. Neither
7. (14, 0), (0, −6)
8. $y = \frac{9}{5}x - 7$; Slope: $\frac{9}{5}$; y-intercept: (0, −7)
9.

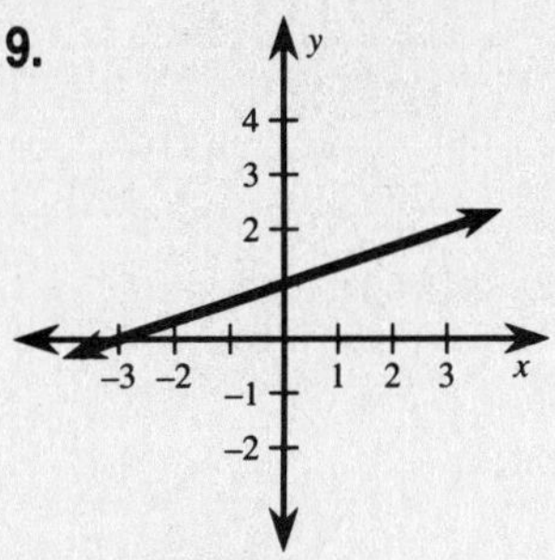

10. $y = -\frac{1}{2}x$
11. $C = 0.20x + 25$
12. No
13.

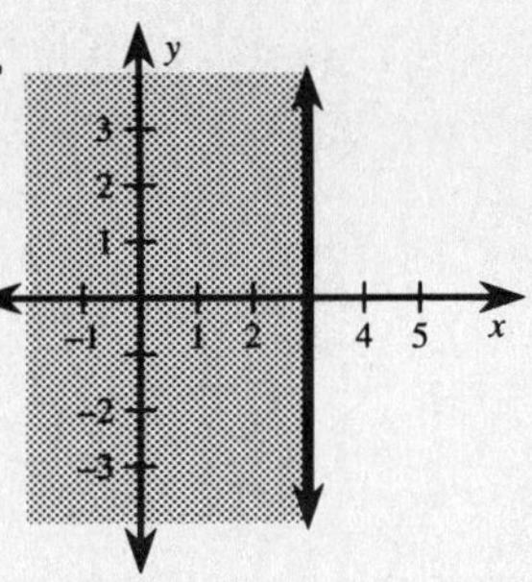

14.

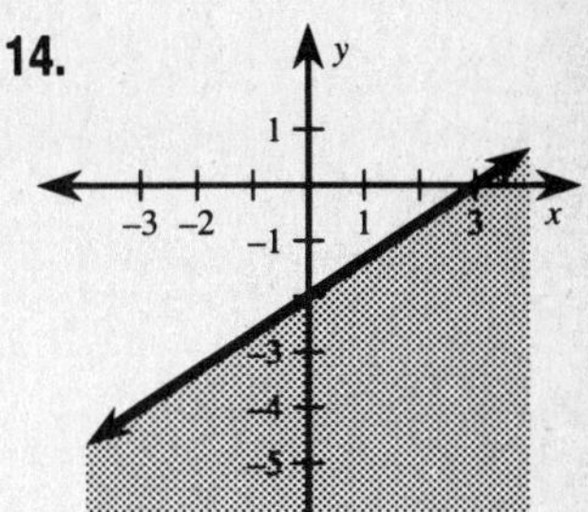

15. (0, 4) 16.

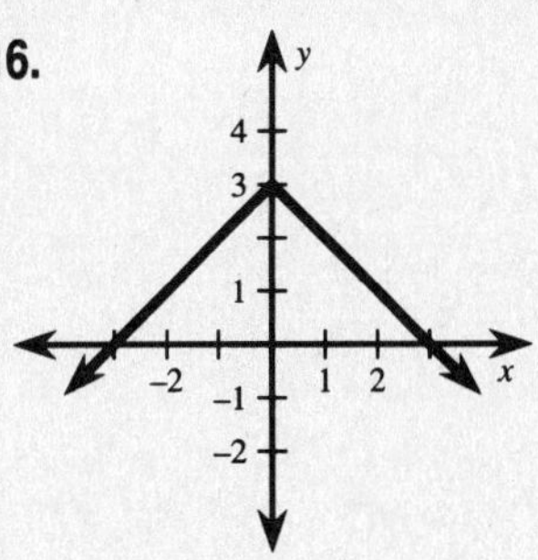

17. 1996, 2000 18. No correlation

Chapter Test 2-C

1. None. It is on the x axis.
2. $x = 0$
3. 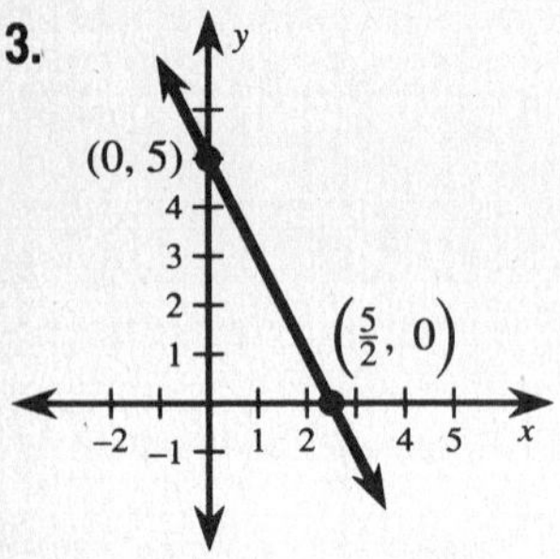

4. $\frac{3}{5}$
5. $y = \frac{7}{4}x + 5$
6. Perpendicular
7. (9, 0), (0, −12)
8. $y = \frac{3}{7}x - \frac{4}{7}$; Slope: $\frac{3}{7}$; y-intercept: $\left(0, -\frac{4}{7}\right)$
9. 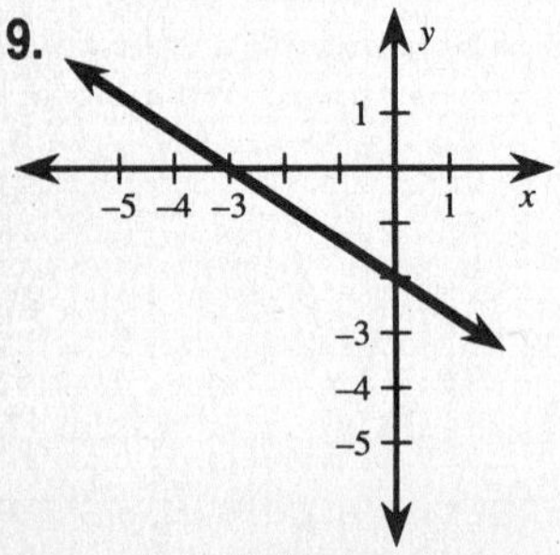
10. $y = -\frac{8}{5}x$
11. $C = 0.25x + 20$
12. Yes
13. 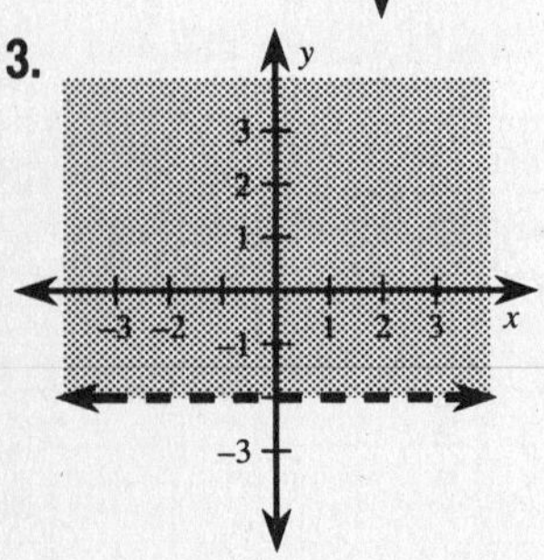
14. 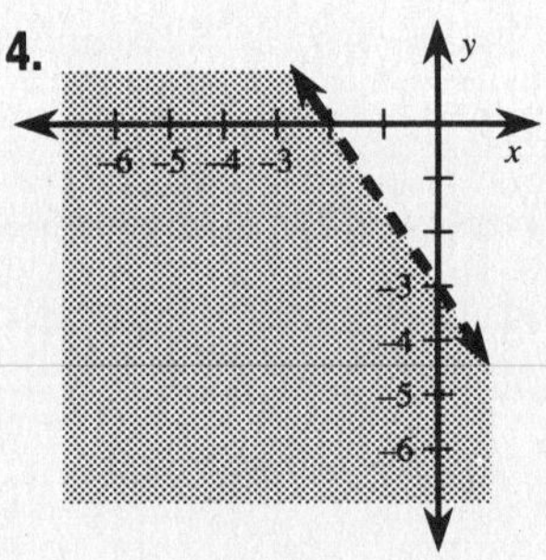
15. Vertex: $\left(\frac{1}{3}, 2\right)$
16. 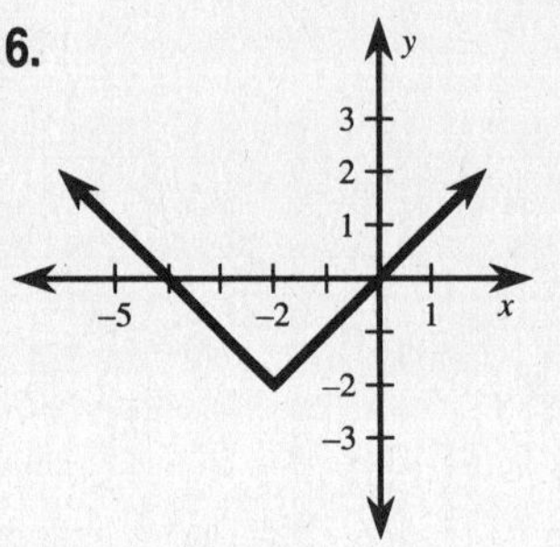
17. 1992, 1998
18. Positive correlation

Mid-Chapter Test 3–A

1. 1
2. No
3. (4, −1)

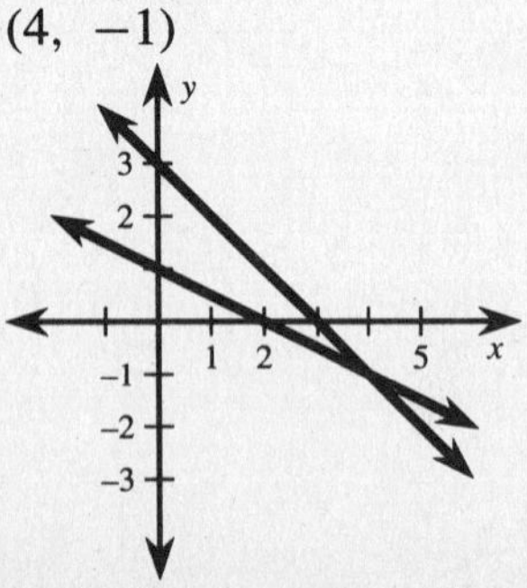

4. (2, 2)
5. $\left(\frac{2}{3}, -\frac{5}{3}\right)$
6. 10 of the \$300 VCRs; 12 of the \$225 VCRs

Mid-Chapter Test 3-B

1. None
2. No
3. (−4, −1)

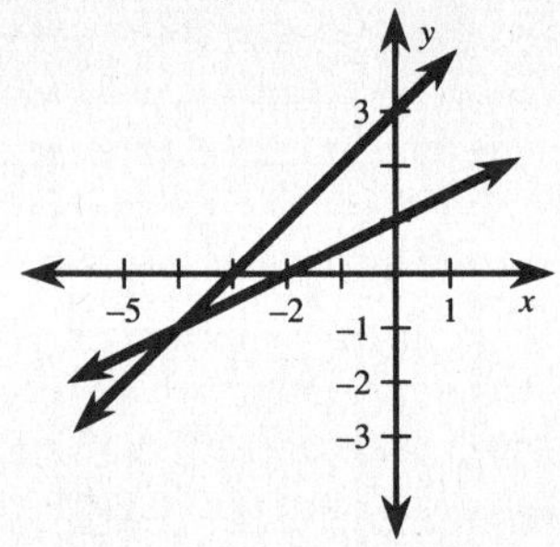

4. (3, −1)
5. $\left(\frac{1}{3}, \frac{3}{2}\right)$
6. 8 of the \$300 VCRs; 16 of the \$225 VCRs

Chapter Test 3-A

1. None
2. Yes
3. (3, 4)

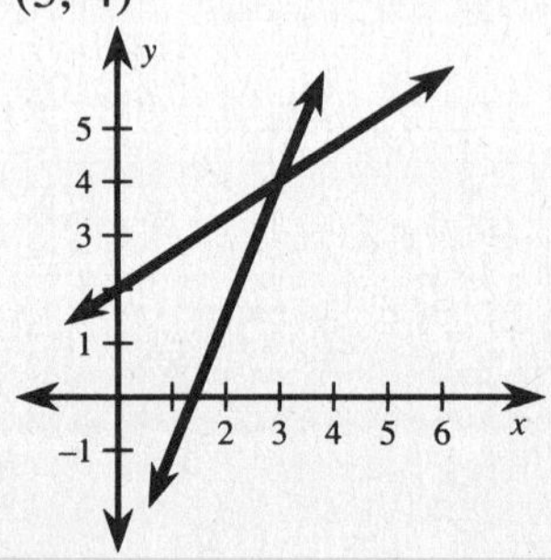

4. 300
5. (3, −8)
6. $\left(-\frac{13}{25}, -\frac{9}{25}\right)$
7. 35.5° and 54.5°
8. 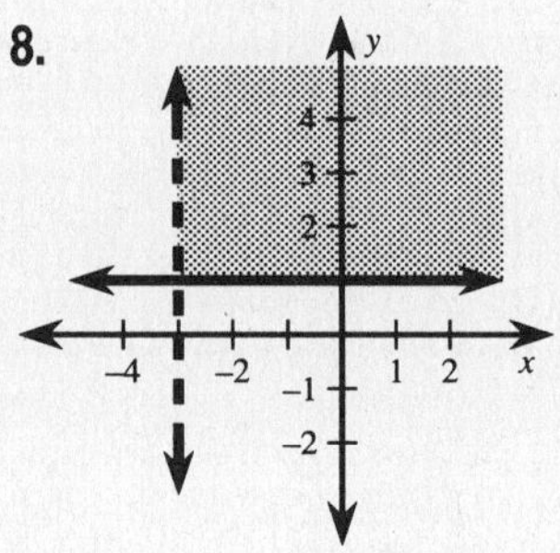
9. 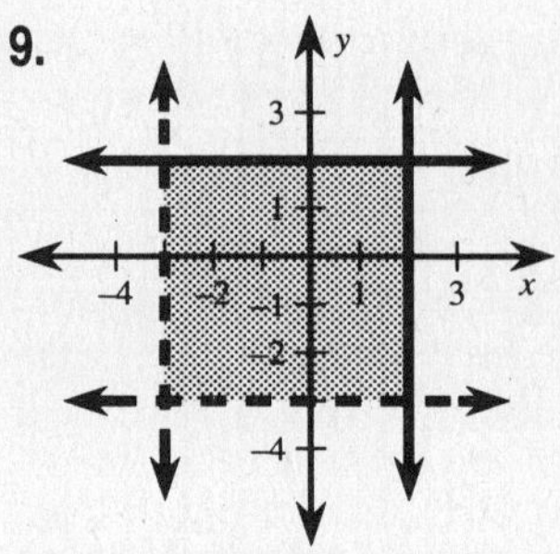
10. 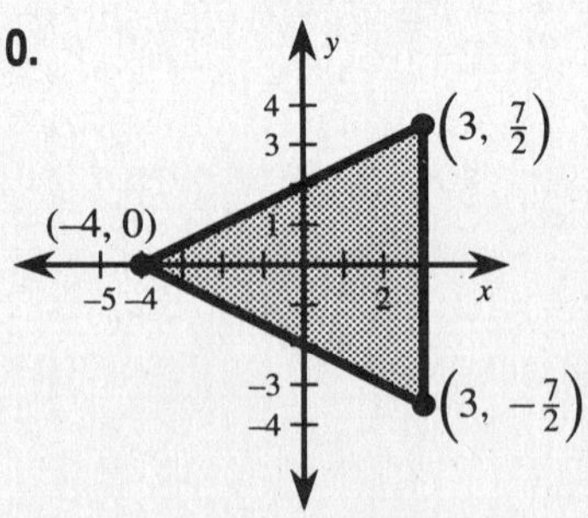

11. 14
12. (2, −1, 3)

Chapter Test 3-B

1. One
2. No
3. $(-1,\ 2)$

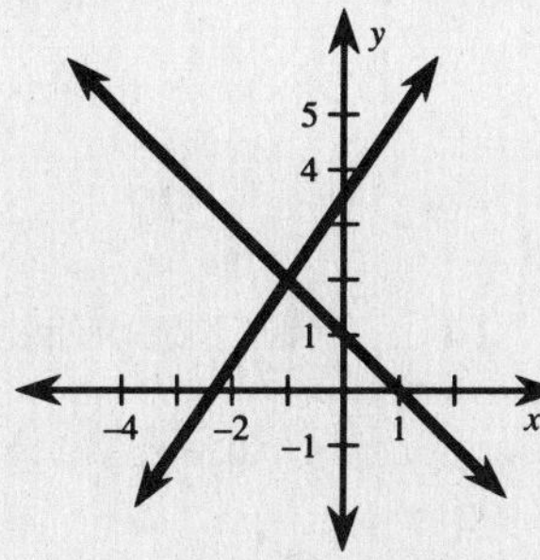

4. 400
5. $(7,\ -6)$
6. $\left(\frac{16}{19},\ -\frac{14}{19}\right)$
7. 37.5° and 52.5°
8.

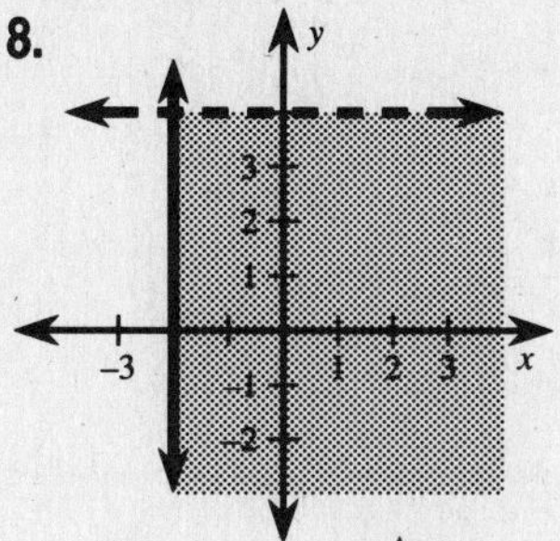

9.

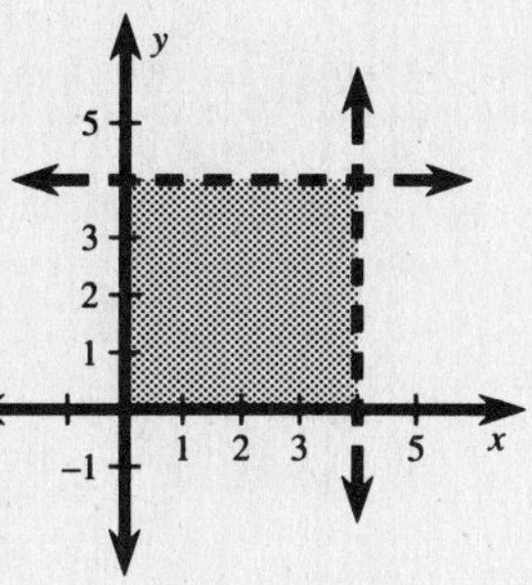

10.

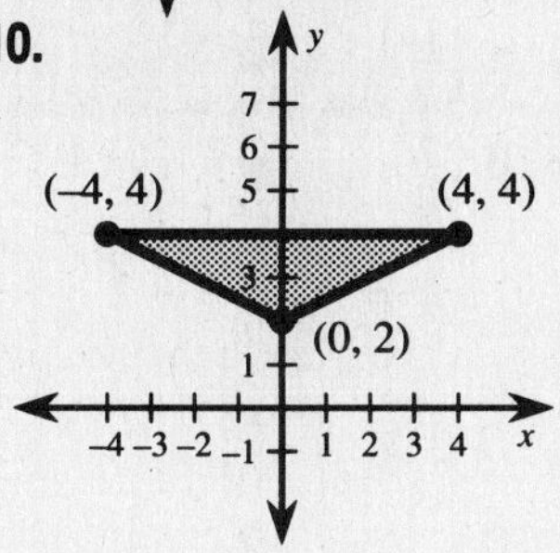

11. -9
12. $(-1,\ 3,\ 2)$

Chapter Test 3-C

1. Many
2. No
3. $(2,\ 3)$

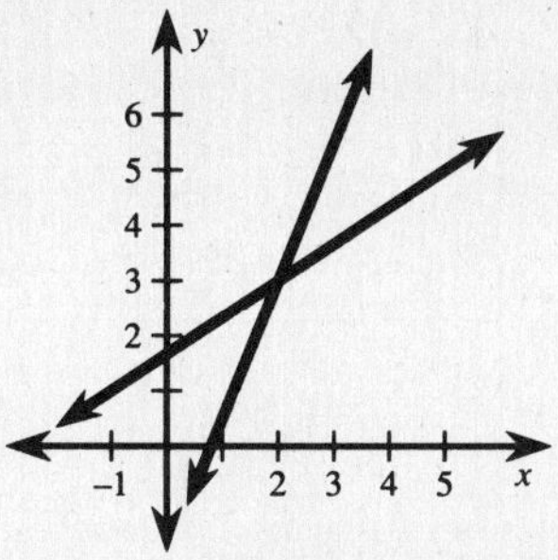

4. 420
5. $\left(\frac{2}{3},\ -\frac{1}{4}\right)$
6. $\left(\frac{20}{13},\ \frac{-4}{13}\right)$
7. 23.5° and 66.5°
8.

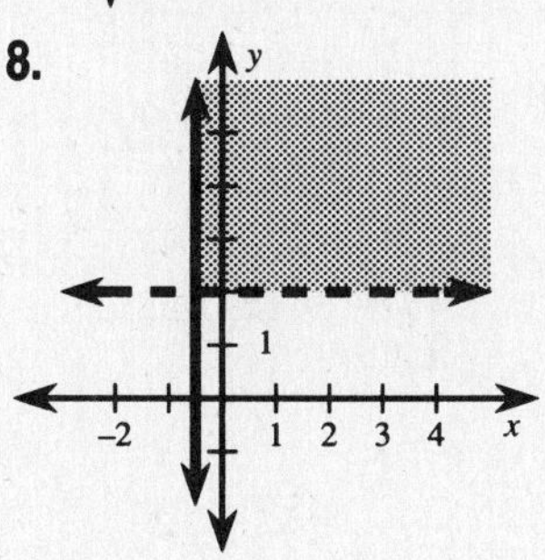

9.

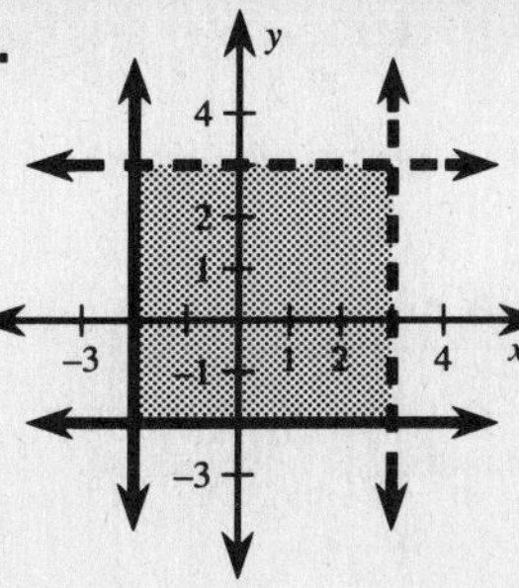

10.

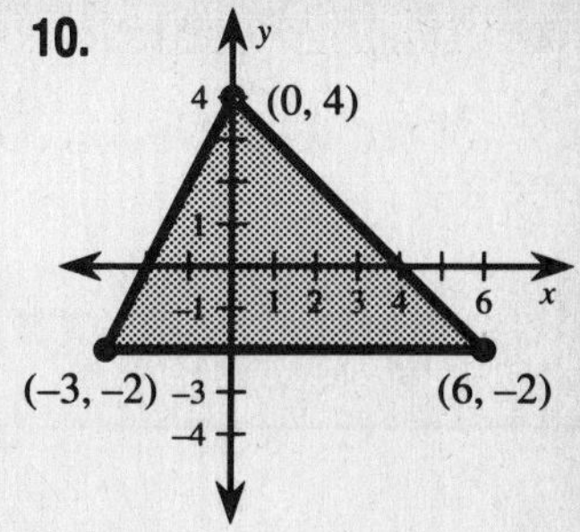

11. Maximum: 10; Minimum: -14
12. $(4,\ -2,\ 1)$

Cumulative Test 1–3

1. $\frac{3}{2}$

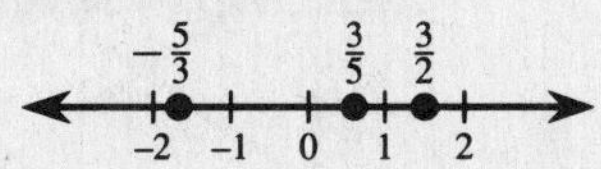

2. $\frac{5}{7},\ \frac{5}{8},\ \frac{3}{5},\ 0,\ -\frac{1}{5},\ -\frac{7}{5},\ -\frac{7}{4}$
3. Associative property of multiplication
4. -10
5. 0
6. 24
7. 91.62
8. $x = 4$
9. $x = \frac{13}{23}$
10. $5^3 y^n$
11. $r = \frac{A-P}{Pt}$
12. 153,000
13. $x > 1$

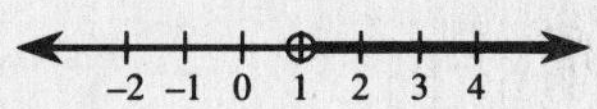

14. Yes
15. $1 \le x \le 4$

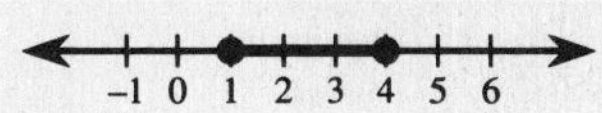

16. $\frac{13}{2},\ 5,\ \frac{7}{2},\ 2,\ \frac{1}{2}\ -1$
17. Slope: $\frac{1}{2}$

    y-intercept: $(0,\ -1)$

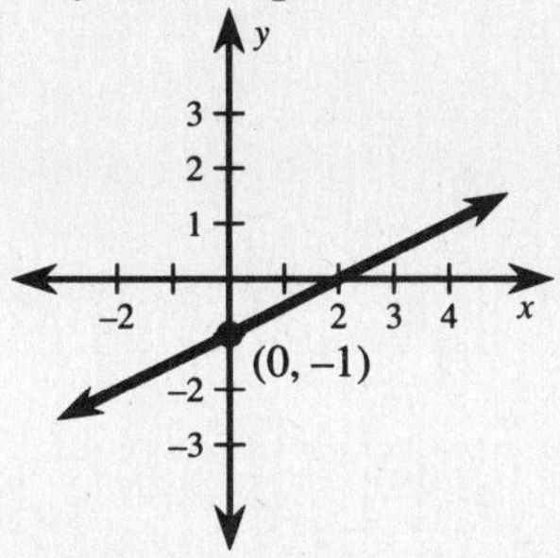

18. $x = 7$
19. $y = -\frac{6}{5}x + 4$
20. $2x + 4y = 5$

    $4x - 2y = 3$
21. $y = -3x + 4$
22. $D = 50 - 22t$
23. $\left(\frac{3}{2},\ 0\right),\ \left(0,\ -\frac{15}{2}\right)$
24. $y = -45x$
25. $C = 0.60t + 0.70,\ t \ge 3$
26. No
27.

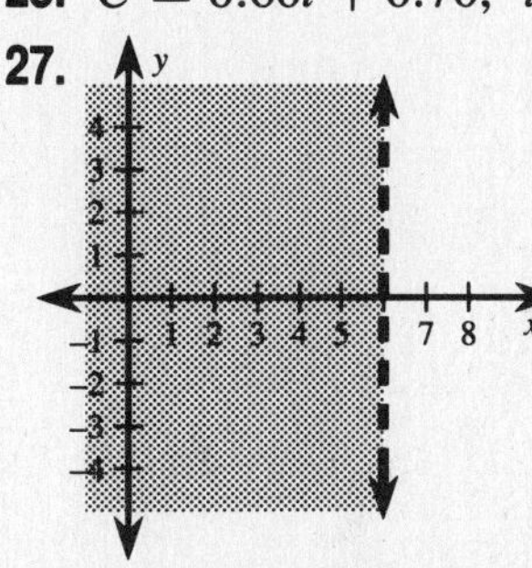

28.

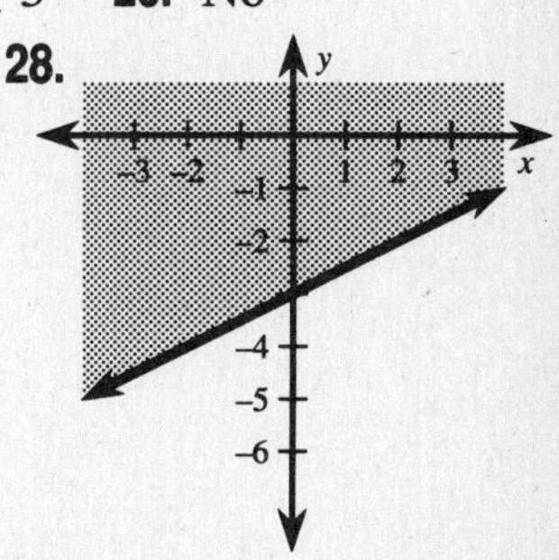

29. 11

30. $(\frac{1}{2}, 5)$

31.

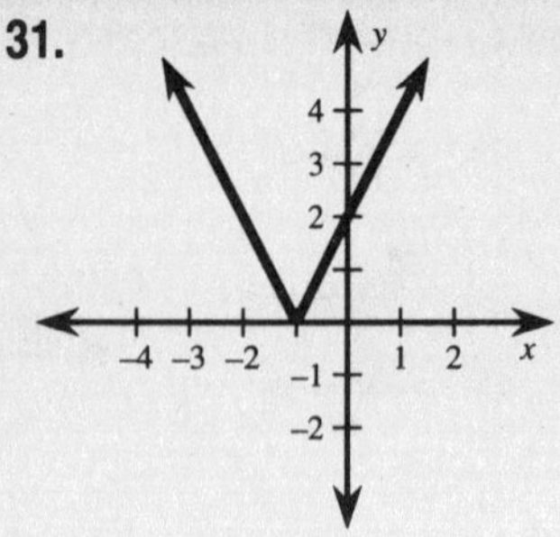

32. Negative correlation

33. None

34. $(3, -1)$

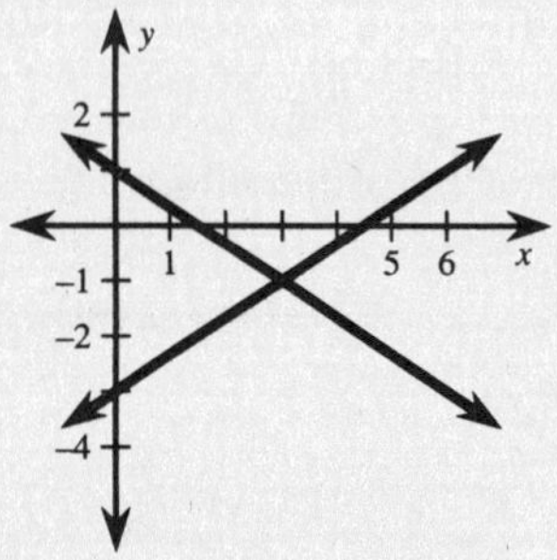

35. 27 dimes, 30 quarters

36. $(5, -2)$ 37. $29.5°$ and $60.5°$

38.

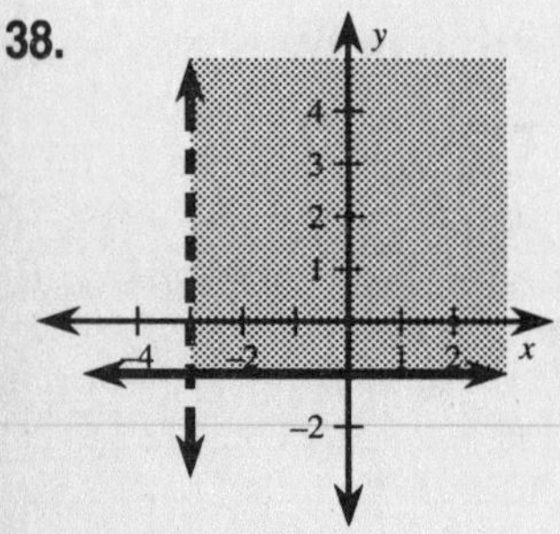

39.

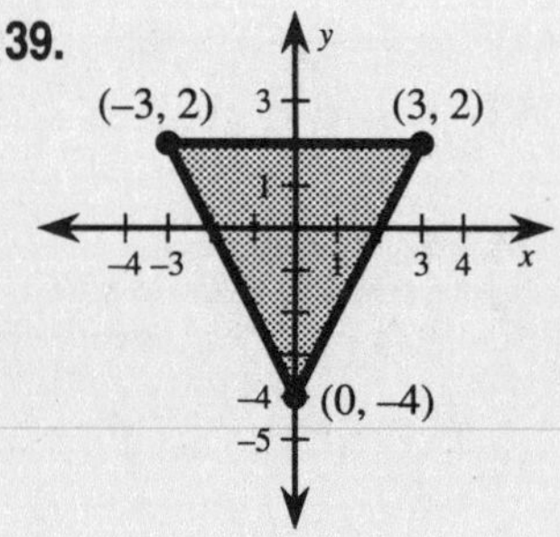

40. $C = 1$

■ Mid-Chapter Test 4-A

1. 2×3 2. $\begin{bmatrix} 4 & 1 & -3 \\ -4 & 6 & 6 \end{bmatrix}$

3. $\begin{bmatrix} 5 & -3 \\ -2 & -2 \end{bmatrix}$ 4. $\begin{bmatrix} -6 & -3 & 15 \\ -3 & 6 & 0 \end{bmatrix}$

5. Not possible 6. 14 7. -26

8. $x = -1, y = -1$ 9. $\begin{bmatrix} \frac{4}{11} & -\frac{3}{11} \\ \frac{1}{11} & \frac{2}{11} \end{bmatrix}$

10. $\frac{225}{2}$

■ Mid-Chapter Test 4-B

1. 2×4 2. $\begin{bmatrix} -3 & 0 & -2 \\ 6 & -2 & -3 \end{bmatrix}$

3. $\begin{bmatrix} 3 & 0 \\ -3 & 12 \end{bmatrix}$ 4. Not possible

5. $\begin{bmatrix} -4 & -4 & 8 \\ -8 & 0 & 4 \end{bmatrix}$ 6. 2 7. 9

8. $\begin{bmatrix} \frac{1}{19} & -\frac{4}{19} \\ \frac{3}{19} & \frac{7}{19} \end{bmatrix}$ 9. $x = 1, y = -1$ 10. 56

■ Chapter Test 4-A

1. 3×2 2. $\begin{bmatrix} 4 & 2 & -7 \\ 10 & 9 & 4 \\ 3 & 12 & 2 \end{bmatrix}$

3. $\begin{bmatrix} -4 & -17 \\ 9 & -4 \end{bmatrix}$ 4. $\begin{bmatrix} 10 & 5 & -25 \\ 30 & -15 & 20 \end{bmatrix}$

5. $\begin{bmatrix} -5 & 14 \\ 27 & 14 \\ 14 & -4 \end{bmatrix}$ 6. Not possible

7. $x = 4, y = -1$ 8. 31 9. -92

10. $\begin{bmatrix} \frac{1}{17} & \frac{2}{17} \\ -\frac{7}{17} & \frac{3}{17} \end{bmatrix}$ 11. $\frac{37}{2}$

12. \$198 at Store 1; \$203 at Store 2

13. $\begin{bmatrix} 5 & -2 \\ 4 & 3 \end{bmatrix}\begin{bmatrix} x \\ y \end{bmatrix} = \begin{bmatrix} 8 \\ 7 \end{bmatrix}$

14. $\begin{bmatrix} -1 & 8 \\ -1 & 7 \end{bmatrix}\begin{bmatrix} -21 \\ -2 \end{bmatrix} = \begin{bmatrix} 5 \\ 7 \end{bmatrix}, x = 5, y = 7$

15. $\left[\begin{array}{ccc:c} 1 & -2 & 1 & 4 \\ 2 & 1 & 0 & 8 \\ 1 & -3 & 2 & -1 \end{array}\right]$

16. $\begin{cases} x + 5y - 7z = 3 \\ y + 3z = -4 \\ z = 7 \end{cases}$

17. $x = 4, y = -2, z = 5$

18. $\begin{bmatrix} 3 & -5 \\ 2 & 4 \end{bmatrix}, \begin{vmatrix} 3 & -5 \\ 2 & 4 \end{vmatrix} = 22$

19. $x = \dfrac{\begin{vmatrix} 0 & 4 \\ -2 & 5 \end{vmatrix}}{\begin{vmatrix} 5 & 4 \\ 6 & 5 \end{vmatrix}} = 8$

$y = \dfrac{\begin{vmatrix} 5 & 0 \\ 6 & -2 \end{vmatrix}}{\begin{vmatrix} 5 & 4 \\ 6 & 5 \end{vmatrix}} = -10$

20. $x = \dfrac{\begin{vmatrix} -4 & 1 & 1 \\ 1 & -1 & 3 \\ 4 & 2 & -1 \end{vmatrix}}{\begin{vmatrix} 1 & 1 & 1 \\ 2 & -1 & 3 \\ 3 & 2 & -1 \end{vmatrix}} = \dfrac{39}{13} = 3$

$y = \dfrac{\begin{vmatrix} 1 & -4 & 1 \\ 2 & 1 & 3 \\ 3 & 4 & -1 \end{vmatrix}}{13} = -\dfrac{52}{13} = -4$

$z = \dfrac{\begin{vmatrix} 1 & 1 & -4 \\ 2 & -1 & 1 \\ 3 & 2 & 4 \end{vmatrix}}{13} = -\dfrac{39}{13} = -3$

■ Chapter Test 4-B

1. 3×4 **2.** $\begin{bmatrix} 10 & -1 & 11 \\ 2 & 4 & 17 \\ 6 & -3 & 8 \end{bmatrix}$

3. $\begin{bmatrix} -4 & 3 \\ 4 & -8 \end{bmatrix}$ **4.** $\begin{bmatrix} -6 & -3 & 12 \\ -3 & 0 & -6 \end{bmatrix}$

5. [4] **6.** Not possible

7. $x = 4,\ y = 2$ **8.** 83 **9.** 15

10. $\begin{bmatrix} -\frac{1}{16} & \frac{7}{48} \\ \frac{3}{16} & -\frac{5}{48} \end{bmatrix}$ **11.** 17

12. \$226 at Store 1; \$182 at Store 2

13. $\begin{bmatrix} 4 & -10 \\ 3 & 9 \end{bmatrix}\begin{bmatrix} x \\ y \end{bmatrix} = \begin{bmatrix} 7 \\ 4 \end{bmatrix}$

14. $\begin{bmatrix} 2 & -1 \\ -5 & 3 \end{bmatrix}\begin{bmatrix} 3 \\ 2 \end{bmatrix} = \begin{bmatrix} 4 \\ -9 \end{bmatrix}$

$x = 4,\ y = -9$

15. $\left[\begin{array}{ccc|c} 1 & -3 & 4 & 2 \\ 2 & 1 & -1 & -1 \\ 1 & 2 & 1 & 3 \end{array}\right]$

16. $\begin{cases} x - 3y + 2z = 7 \\ y + 5z = -2 \\ z = 8 \end{cases}$

17. $x = 3,\ y = -2,\ z = 6$

18. $\begin{bmatrix} 8 & -5 \\ 12 & 7 \end{bmatrix}, \begin{vmatrix} 8 & -5 \\ 12 & 7 \end{vmatrix} = 116$

19. $x = \dfrac{\begin{vmatrix} 1 & 1 \\ 5 & 2 \end{vmatrix}}{\begin{vmatrix} 2 & 1 \\ 5 & 2 \end{vmatrix}} = \dfrac{-3}{-1} = 3$

$y = \dfrac{\begin{vmatrix} 2 & 1 \\ 5 & 5 \end{vmatrix}}{-1} = \dfrac{5}{-1} = -5$

20. $x = \dfrac{\begin{vmatrix} -3 & -1 & 1 \\ 15 & 2 & -1 \\ 0 & 3 & 1 \end{vmatrix}}{\begin{vmatrix} 1 & -1 & 1 \\ 3 & 2 & -1 \\ 2 & 3 & 1 \end{vmatrix}} = \dfrac{45}{15} = 3$

$y = \dfrac{\begin{vmatrix} 1 & -3 & 1 \\ 3 & 15 & -1 \\ 2 & 0 & 1 \end{vmatrix}}{15} = \dfrac{0}{15} = 0$

$z = \dfrac{\begin{vmatrix} 1 & -1 & -3 \\ 3 & 2 & 15 \\ 2 & 3 & 0 \end{vmatrix}}{15} = -\dfrac{90}{15} = -6$

■ Chapter Test 4-C

1. 1×5 **2.** $\begin{bmatrix} 3 & 9 & 6 \\ 5 & -6 & 6 \\ 3 & -3 & 5 \end{bmatrix}$

3. $\begin{bmatrix} 5 & 9 & 1 \\ 1 & 4 & -7 \end{bmatrix}$ **4.** $\begin{bmatrix} -8 & 0 & -48 \\ -16 & -32 & 24 \end{bmatrix}$

5. Not possible **6.** $\begin{bmatrix} -2 & -7 \\ 27 & 8 \end{bmatrix}$

7. $x = 4,\ y = 3$ **8.** 155 **9.** 77

10. $\begin{bmatrix} -\frac{8}{31} & \frac{7}{31} \\ \frac{9}{31} & -\frac{4}{31} \end{bmatrix}$ **11.** $\frac{145}{2}$

12. \$600 at Store 1; \$433 at Store 2

13. $\begin{bmatrix} 12 & -37 \\ 9 & 19 \end{bmatrix}\begin{bmatrix} x \\ y \end{bmatrix} = \begin{bmatrix} 14 \\ -7 \end{bmatrix}$

14. $\dfrac{1}{13}\begin{bmatrix} 4 & 5 \\ -1 & 2 \end{bmatrix}\begin{bmatrix} 7 \\ -5 \end{bmatrix} = \dfrac{1}{13}\begin{bmatrix} 3 \\ -17 \end{bmatrix}.$

$x = \frac{3}{13},\ y = -\frac{17}{13}$

15. $\left[\begin{array}{ccc|c} 1 & -4 & 2 & 3 \\ 5 & -7 & 1 & 4 \\ 3 & 2 & -2 & 8 \end{array}\right]$

16. $\begin{cases} x - 3y + 7z = 10 \\ y - 4z = 8 \\ z = -12 \end{cases}$

17. $x = -4,\ y = -1,\ z = 4$

18. $\begin{bmatrix} 9 & -2 \\ 3 & 6 \end{bmatrix}, \begin{vmatrix} 9 & -2 \\ 3 & 6 \end{vmatrix} = 60$

19. $x = \dfrac{\begin{vmatrix} -2 & -5 \\ 5 & 3 \end{vmatrix}}{\begin{vmatrix} 3 & -5 \\ 4 & 3 \end{vmatrix}} = \frac{19}{29}$

$y = \dfrac{\begin{vmatrix} 3 & -2 \\ 4 & 5 \end{vmatrix}}{29} = \frac{23}{29}$

20. $x = \dfrac{\begin{vmatrix} 0 & 2 & 0 \\ 3 & -1 & 2 \\ 0 & 5 & 6 \end{vmatrix}}{\begin{vmatrix} 3 & 2 & 0 \\ 2 & -1 & 2 \\ 0 & 5 & 6 \end{vmatrix}} = \dfrac{-36}{-72} = \dfrac{1}{2}$

$y = \dfrac{\begin{vmatrix} 3 & 0 & 0 \\ 2 & 3 & 2 \\ 0 & 0 & 6 \end{vmatrix}}{-72} = \dfrac{54}{-72} = -\dfrac{3}{4}$

$z = \dfrac{\begin{vmatrix} 3 & 2 & 0 \\ 2 & -1 & 3 \\ 0 & 5 & 0 \end{vmatrix}}{-72} = \dfrac{-45}{-72} = \dfrac{5}{8}$

Mid-Chapter Test 5-A

1. $\pm\sqrt{5}$ **2.** 41.53
3. Vertex: (1, 6); Axis: $x = 1$; Opens down
4. 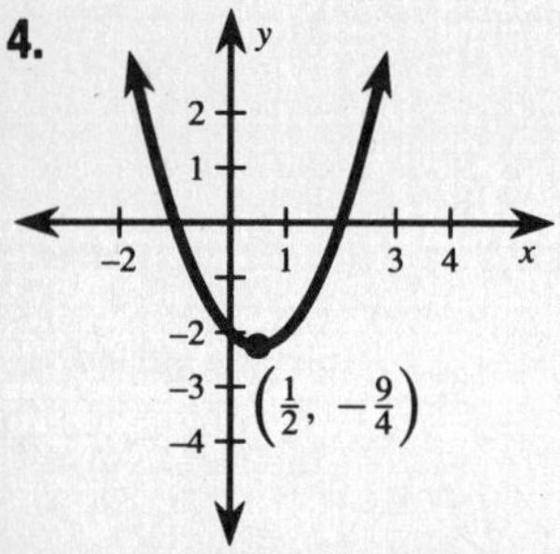

5. $2 \pm \sqrt{7}$
6. Base: 6 in.
   Height: 4 in.
7. 4400 ft

Mid-Chapter Test 5-B

1. $\pm\sqrt{3}$ **2.** 55.93
3. Vertex: (2, 4); Axis: $x = 2$; Opens down
4. 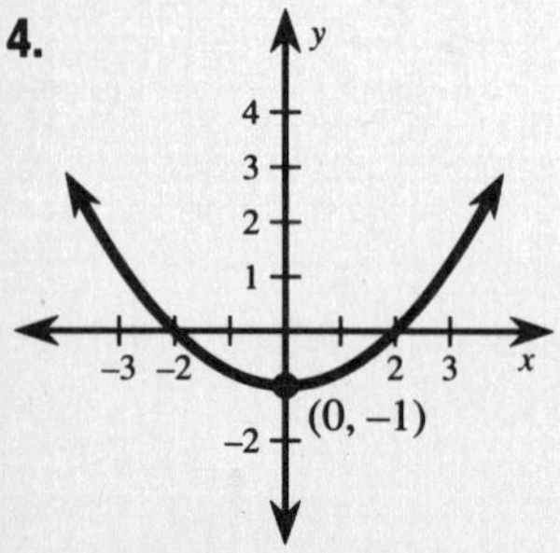

5. $3 \pm \sqrt{10}$
6. Width: 5 in.
   Length: 9 in.
7. 5696 ft

Chapter Test 5-A

1. $x = \pm\sqrt{3}$ **2.** ± 1.34 **3.** 9.45 in.
4. 5.77 seconds **5.** Down
6. Vertex: $(-2, -3)$; Axis: $x = -2$
7. 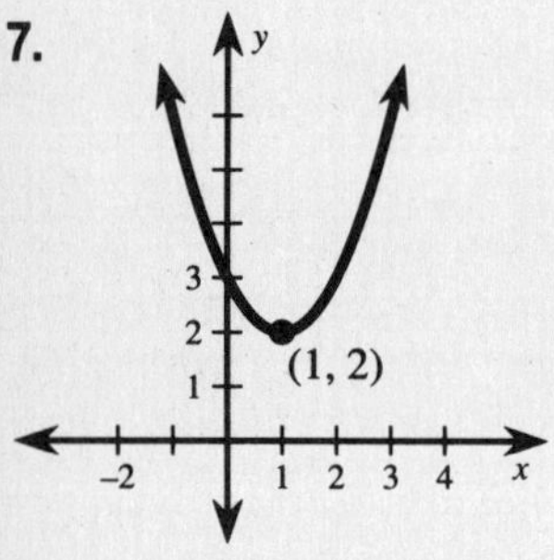

8. $(x - 9)^2$
9. 5, -7
10. Height: 10 ft
    Base: 7 ft
11. 249
12. 2
13. $-1 \pm \sqrt{2}$ **14.** 0.39, -1.94 **15.** $6i$
16. -9
17. Yes
18. $\pm 2i$
19. $-1 - 4i$
20. $14 - 5i$
21. $-\frac{1}{4} \pm \frac{i\sqrt{23}}{4}$
22. 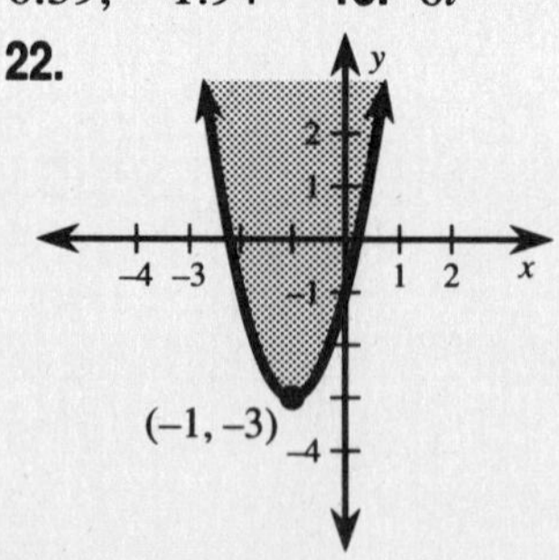

Chapter Test 5-B

1. $\pm\sqrt{7}$ **2.** ± 1.89 **3.** 7.96 in.
4. 4.56 seconds **5.** Up
6. Vertex: (2, 4); Axis: $x = 2$
7. 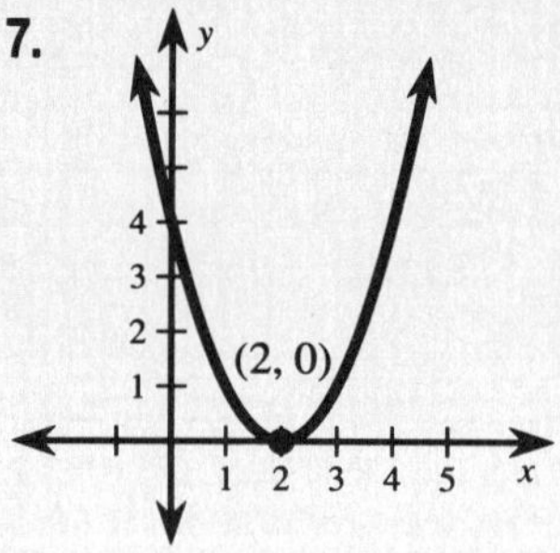

8. $\left(x - \frac{1}{3}\right)^2$
9. -6, 4
10. Base: 8 ft;
    Height: 11 ft
11. -47
12. None
13. $-\frac{1}{6} \pm \frac{\sqrt{13}}{6}$ **14.** 2.08, -1.60 **15.** $7i$
16. -3
17. No
18. $\pm i\sqrt{5}$
19. $8 - 4i$
20. $11 + 13i$
21. $\frac{1}{4} \pm i\frac{\sqrt{15}}{4}$
22. 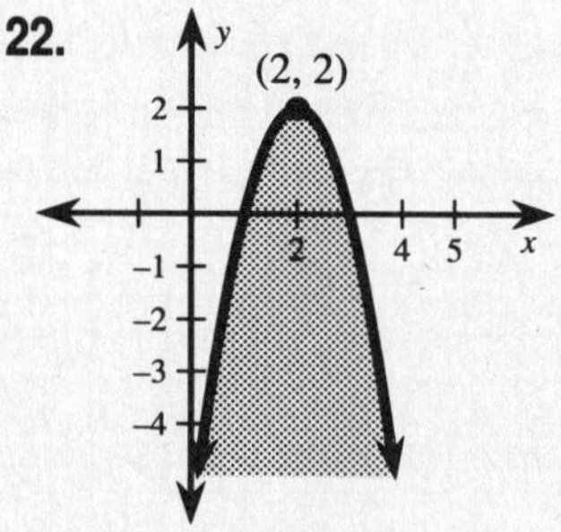

Chapter Test 5-C

1. $\pm 4\sqrt{6}$ **2.** ± 3.11
3. 7.72 centimeters **4.** 3.54 seconds
5. Vertex: $(-2, 15)$; Opens down
6. Axis: $x = 1$; Intercepts: (0, 0), (2, 0)
7. 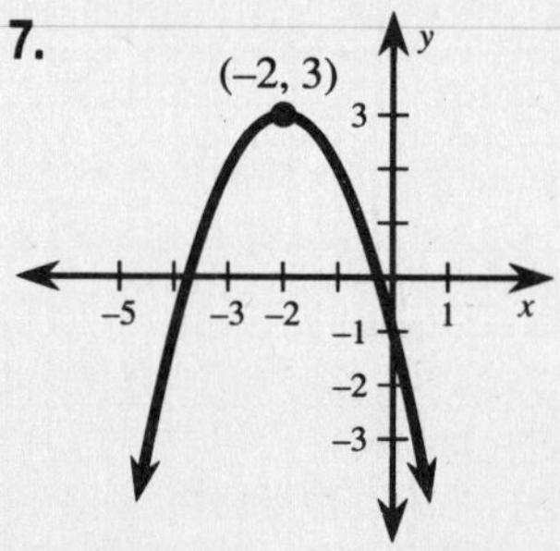

8. $\left(x - \frac{3}{5}\right)^2$
9. $x = 2 \pm \sqrt{6}$
10. Base: 15 ft
    Height: 10 ft
11. 169
12. One **13.** $-\frac{1}{2} \pm \frac{\sqrt{33}}{2}$ **14.** 0.06, -0.16
15. $-9i$ **16.** $3i$ **17.** Yes
18. $\pm i\sqrt{3}$
19. $1 + 3i$
20. $6 + 10i$
21. $\frac{1}{2} \pm i\frac{\sqrt{7}}{2}$
22. 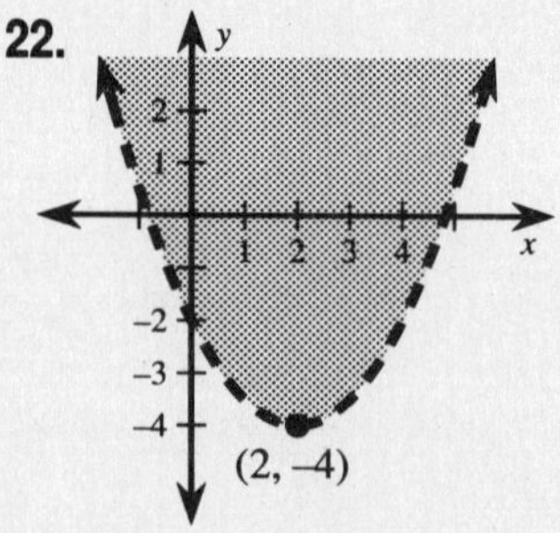

Mid-Chapter Test 6-A

1. It is. **2.** -2

3. Domain: $x \geq -1$; Range: $f(x) \geq 0$

4. $-2x^2 + 5x - 2$ **5.** $3 - 2x$

6. $f(x) = 1.06x$

7. $y = \frac{6-x}{7}$ **8.** No

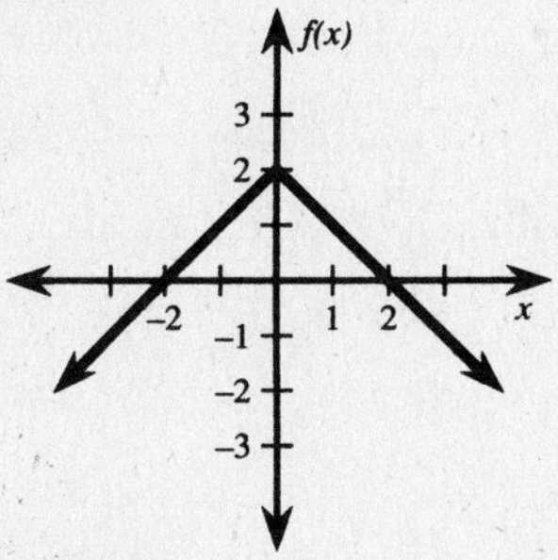

Mid-Chapter Test 6-B

1. It is not. **2.** -20

3. Domain: $x \leq 3$; Range: $f(x) \leq 0$

4. $5x - 7$ **5.** $-2 - 3x$

6. $f(x) = 1.07x$

7. $y = \frac{x+8}{13}$ **8.** No

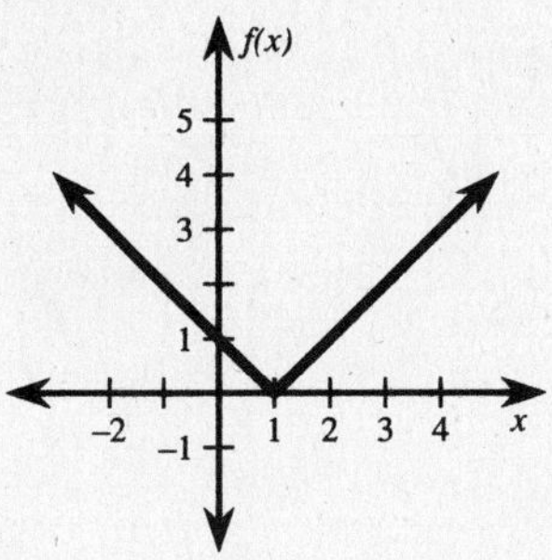

Chapter Test 6-A

1. It is. **2.** -3 **3.** 9π

4. Domain: $x > 4$; Range: $f(x) > 0$

5. $h(x) = \frac{4x+1}{3-2x}$ **6.** $13 - 8x$

7. All real numbers x such that $x \neq \frac{3}{2}$

8. (1, 4), (3, 3), (5, 2), (7, 1)

9. $y = \frac{9-x}{11}$ **10.** They are.

11. No

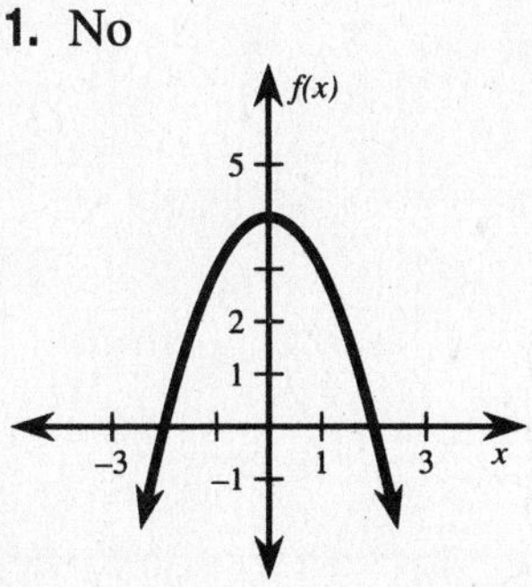

12.

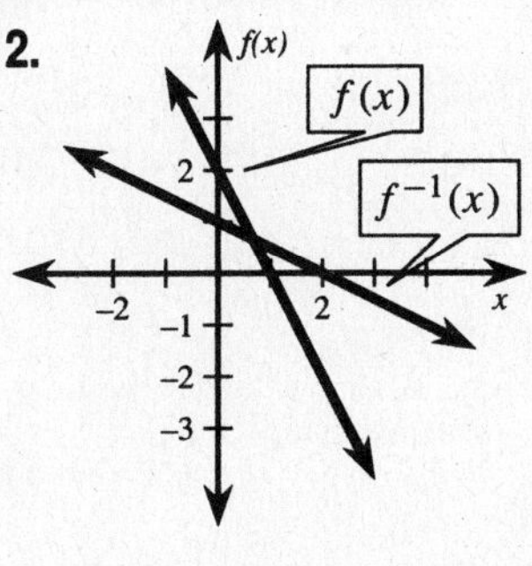

13. -15

14. $f(x) = \begin{cases} -x + 4, & x \leq 4 \\ x - 4, & x > 4 \end{cases}$

or

$f(x) = \begin{cases} -x + 4, & x < 4 \\ x - 4, & x \geq 4 \end{cases}$

15.

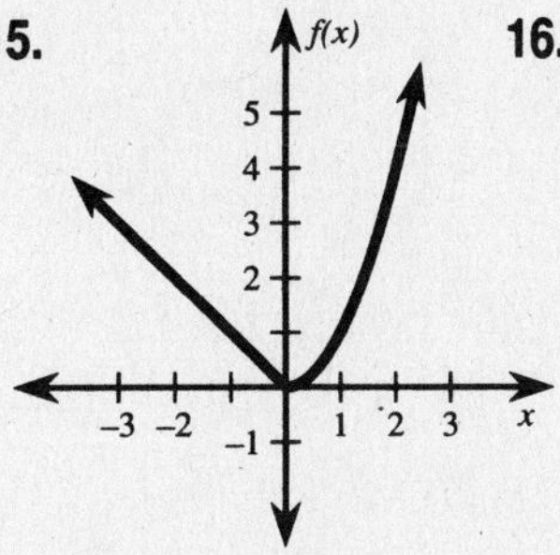

16.

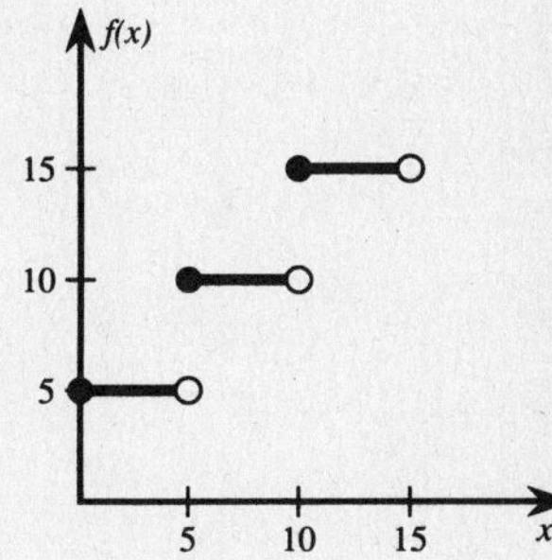

17. Shift the graph of g 5 units down

18.

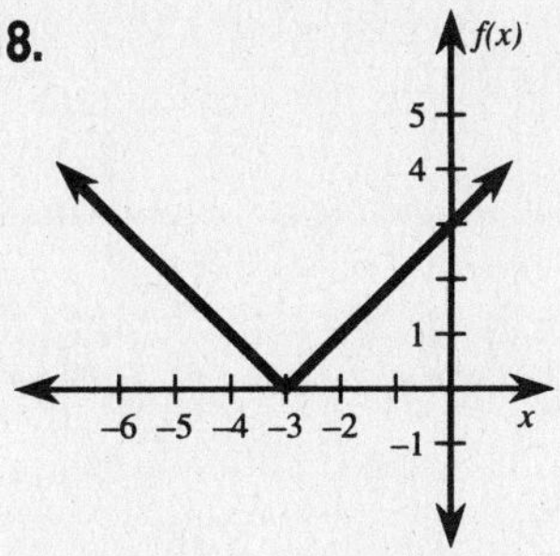

19. 5, 3, 0, -4, -9

20. Mean: 38
Median: 36
Mode: 29

Chapter Test 6-B

1. It is not. **2.** 13 **3.** $\frac{16\pi}{9}$

4. Domain: $x > 3$; Range: $f(x) < 0$

5. $h(x) = \frac{3-2x}{4x+1}$ **6.** $1 - 8x$

7. All real numbers x such that $x \neq -\frac{1}{4}$

8. (5, 1), (3, 2), (1, 3), $(-1, 4)$

9. $y = \frac{x+7}{18}$ **10.** They are.

11. No

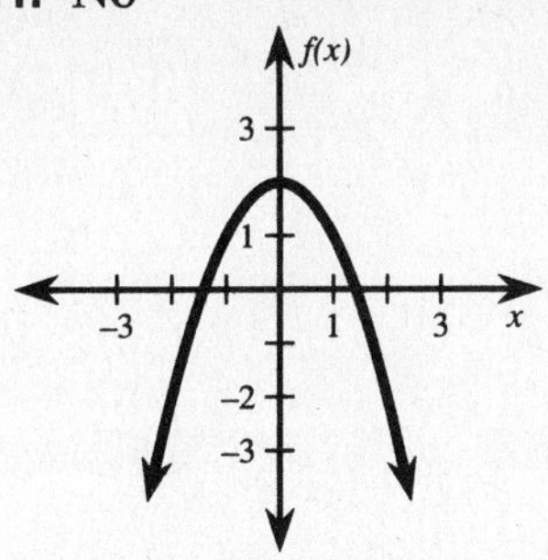

12.

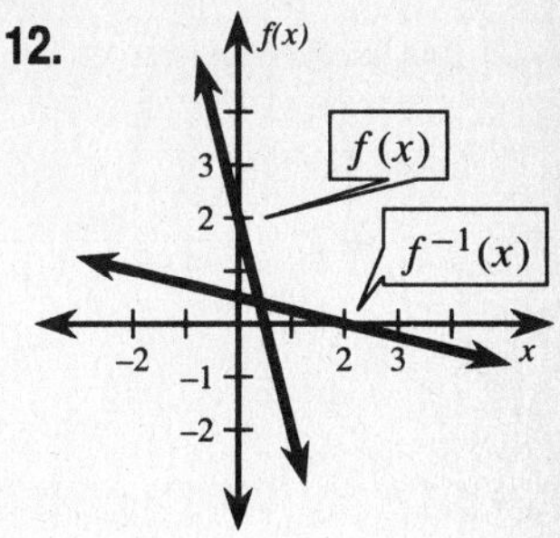

13. -3

14. $f(x) = \begin{cases} -2x - 3, & x \leq -\frac{3}{2} \\ 2x + 3, & x > -\frac{3}{2} \end{cases}$

or

$f(x) = \begin{cases} -2x - 3, & x < -\frac{3}{2} \\ 2x + 3, & x \geq -\frac{3}{2} \end{cases}$

15. 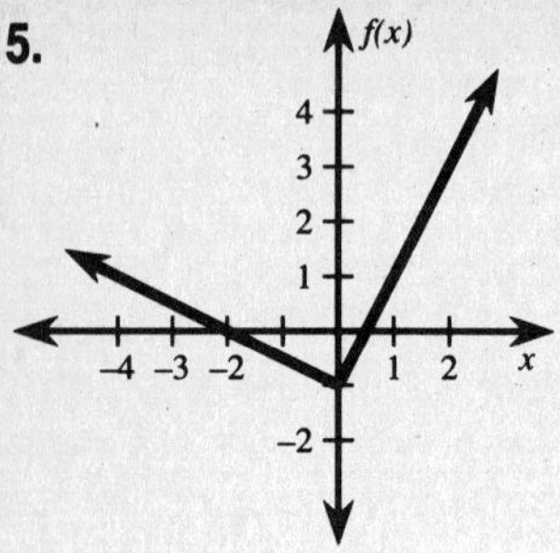

16. 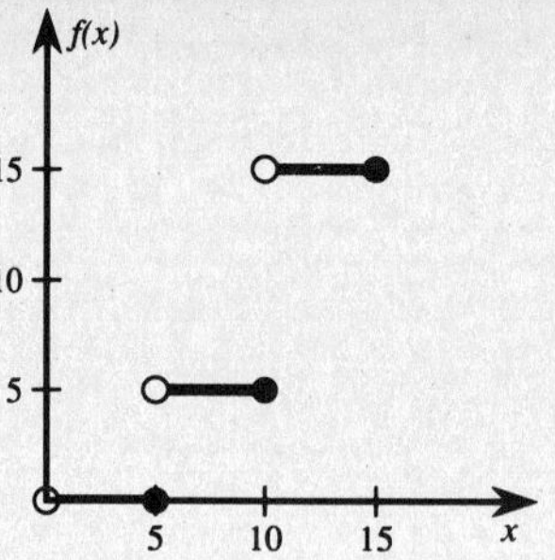

17. Shift the graph of g 4 units upward

18. 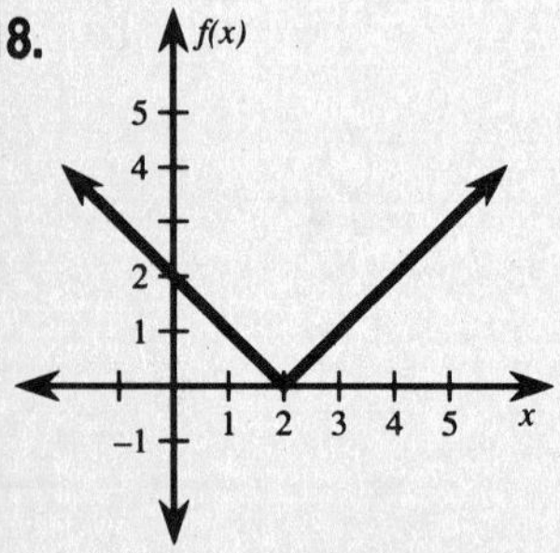

19. 5, 7, 10, 14, 19

20. Mean: 66.7
Median: 64
Mode: 77

Chapter Test 6-C

1. It is. 2. 3 3. $\frac{9\pi}{2}$
4. Domain: $x \geq -1, x \neq 0$; Range: $f(x) \geq 0$
5. $h(x) = \frac{x^2+4}{2x-3}$ 6. $2x^2 + 5$
7. All real numbers x such that $x \neq \frac{3}{2}$
8. (1, 1), (2, 2), (3, 1), (4, 4)
9. $y = \frac{x+7}{-22}$ 10. No
11. No

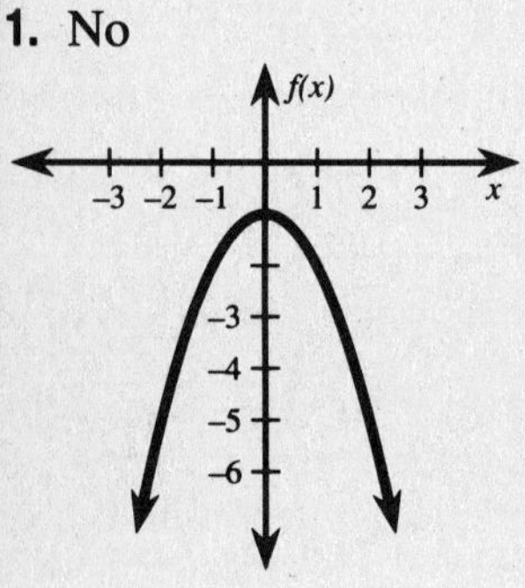

12. 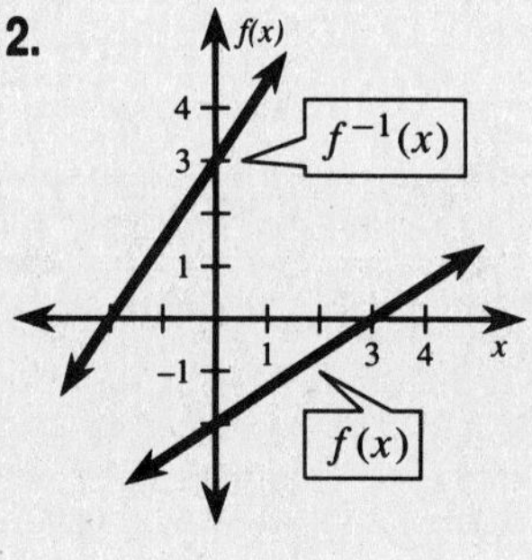

13. 3 14. $f(x) = \begin{cases} 3 - 5x, & x \leq \frac{3}{5} \\ -3 + 5x, & x > \frac{3}{5} \end{cases}$

or

$f(x) = \begin{cases} 3 - 5x, & x < \frac{3}{5} \\ -3 + 5x, & x \geq \frac{3}{5} \end{cases}$

15. 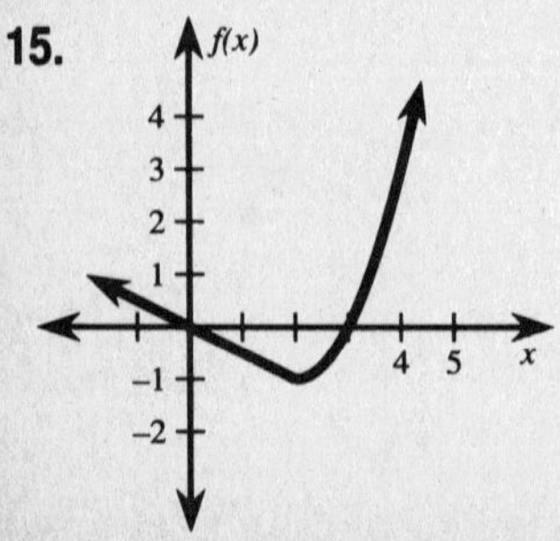

16. 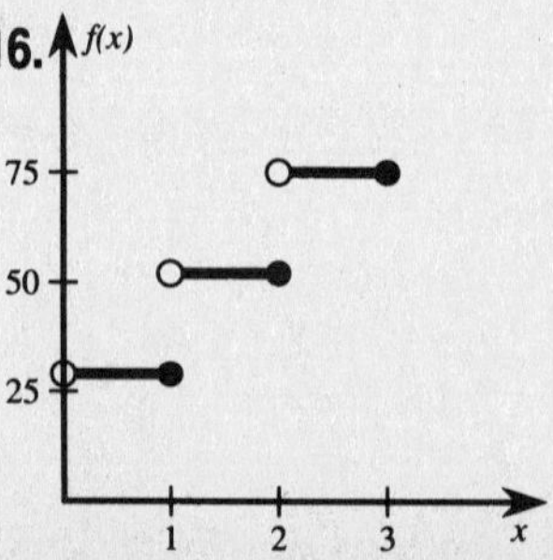

17. Shift the graph of g 2 units to the left

18. 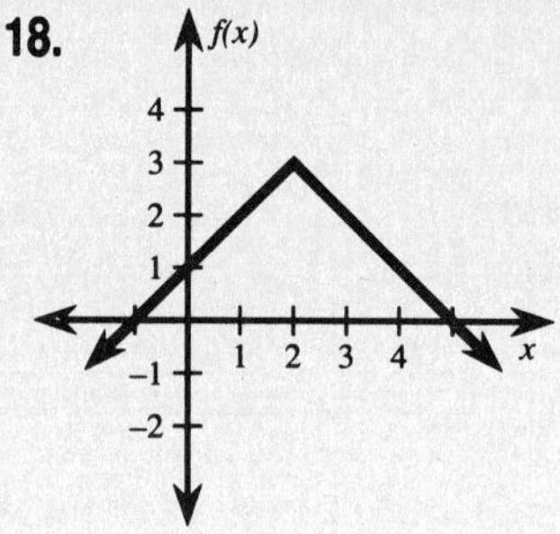

19. 5, −3, 6, −2, 7

20. Mean: 91.9
Median: 91.5
Mode: 90

Cumulative Test 1–6

1. $-\frac{3}{2}, -\frac{2}{3}, -\frac{3}{5}, 0, \frac{4}{7}, \frac{5}{3}, \frac{7}{4}$
2. Commutative property of multiplication
3. 4 4. 37.72
5. $x = -\frac{7}{2}$ 6. $t = \frac{A-P}{rP}$
7. $x > 1$ 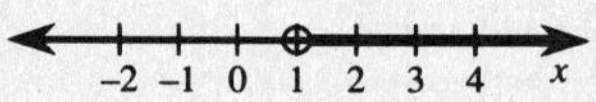
8. $-1 \leq x \leq 4$ 
9. $y = -2$ 10. $y = \frac{4}{7}x - 3$
11. $x - 2y = 5$ and $2x - 4y = 5$
12. $D = 15 - 2t$
13. $y = \frac{9}{2}x$
14. No

15. 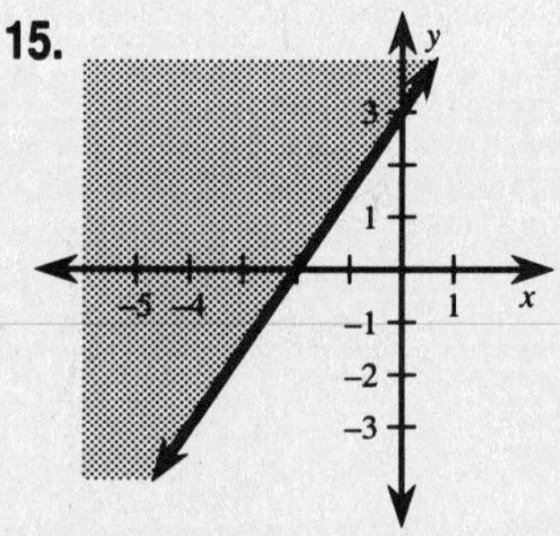

16. −13 17. 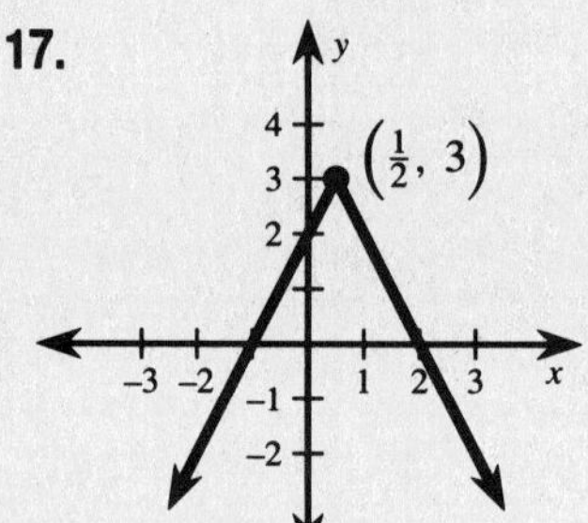

18. Many 19. $(-\frac{4}{7}, \frac{20}{21})$

20. 9

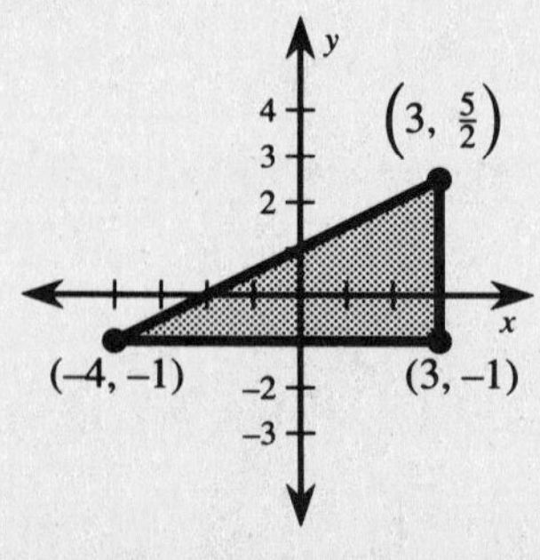

21. $\begin{bmatrix} -4 & 14 \\ -17 & 14 \end{bmatrix}$

22. $\begin{bmatrix} -9 & 11 \\ 9 & 15 \end{bmatrix}$

23. $x = -2$
$y = -3$

24. $\begin{bmatrix} \frac{1}{17} & \frac{7}{68} \\ -\frac{2}{17} & \frac{3}{68} \end{bmatrix}$ 25. 36 26. 9

27. $x = 2, y = -1, z = -3$

28. $x = \dfrac{\begin{vmatrix} 8 & -4 \\ -3 & 7 \end{vmatrix}}{\begin{vmatrix} 5 & -4 \\ 2 & 7 \end{vmatrix}} = \frac{44}{43}$, $y = \dfrac{\begin{vmatrix} 5 & 8 \\ 2 & -3 \end{vmatrix}}{43} = -\frac{31}{43}$

$\left(\frac{44}{43}, -\frac{31}{43}\right)$

29. $\pm\sqrt{6}$ 30. $\frac{-5\pm\sqrt{37}}{6}$

31.

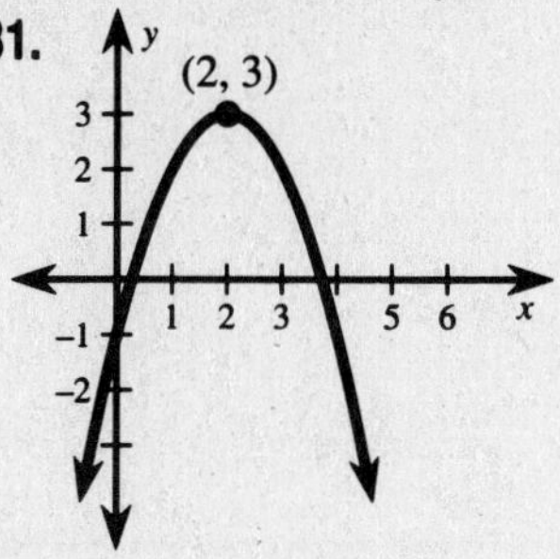

32. 1.12, −0.42

33. $-4 + 3i$

34. $8 - 16i$

35. $-\frac{1}{8} \pm i\frac{\sqrt{15}}{8}$

36. It is. Inverse: (4, −2), (3, −1), (2, 1), (1, 2)

37. $f(x) \cdot g(x) = -x^2 - \frac{8}{3}x + 1$

$f(g(x)) = -x - 2$

38. $f^{-1}(x) = -3x - 3$ 39. 4

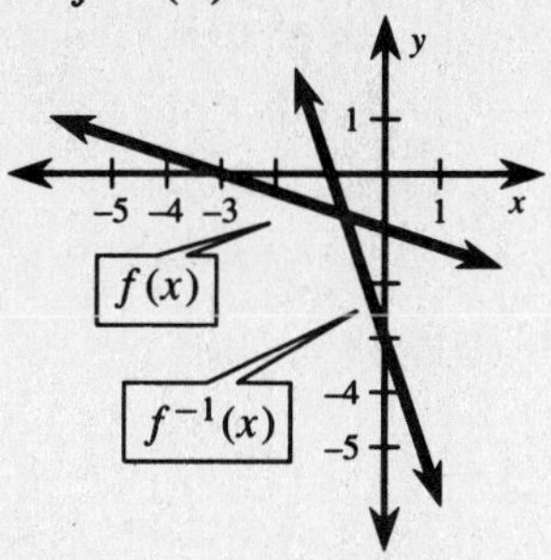

40. $f(x) = \begin{cases} -3x + 6, & x \le 2 \\ 3x - 6, & x > 2 \end{cases}$

or

$f(x) = \begin{cases} -3x + 6, & x < 2 \\ 3x - 6, & x \ge 2 \end{cases}$

41. Shift the graph of g one unit to the right

42. 3, 7, 13, 21

43. Mean: 8.36; Median: 8; Mode: 7

Mid-Chapter Test 7-A

1. $\frac{y}{x^3}$ 2. −1 3. \$6241.82

4. Decay 5. $20^{1/7}$ 6. $\pm\sqrt[4]{23}$

7. 3.187 8. 59%

Mid-Chapter Test 7-B

1. x^3y 2. $\frac{3}{2}$ 3. \$4510.49

4. Growth 5. $17^{1/6}$ 6. $\pm\sqrt[4]{77}$

7. 3.380 8. 53%

Chapter Test 7-A

1. $-\frac{27}{x^6}$ 2. $\frac{2x^3}{3y^5}$ 3. $x = 5$

4. \$1967.15 5. \$452.89 6. \$16,734.32

7. 32 8. 0.481 9. 5.07 cm

10. $\sqrt[5]{7}$ 11. 4.340 12. $\frac{1}{5}$

13. $6\sqrt[3]{5}$ 14. $5^{1/6}$ 15. $y = 127$

16. 5

17. 5

18. $x \ge -1$

19. $g(x) \ge 2$

20.

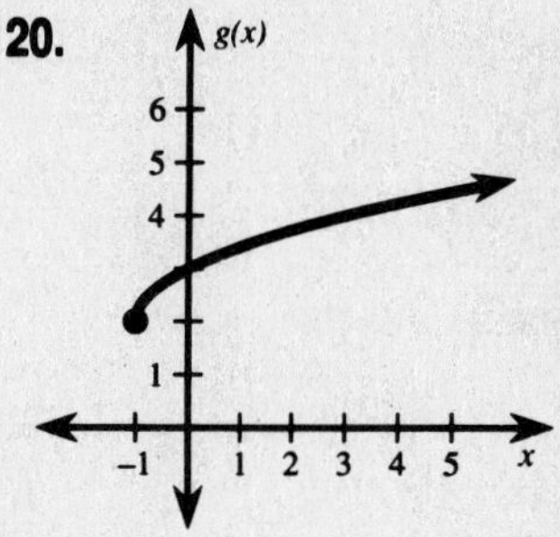

Chapter Test 7-B

1. $\frac{4}{x^6}$ 2. $\frac{x^2}{2y^5}$ 3. 3

4. \$1872.98 5. \$666.34 6. \$26,972.59

7. 9 8. 0.397 9. 4.28 cm

10. $\sqrt[6]{18}$ 11. 3.375 12. 7

13. $7\sqrt[4]{2}$ 14. $7^{1/6}$ 15. $y = 64$

16. 4

17. 10

18. $x \ge -3$

19. $g(x) \ge 1$

20.

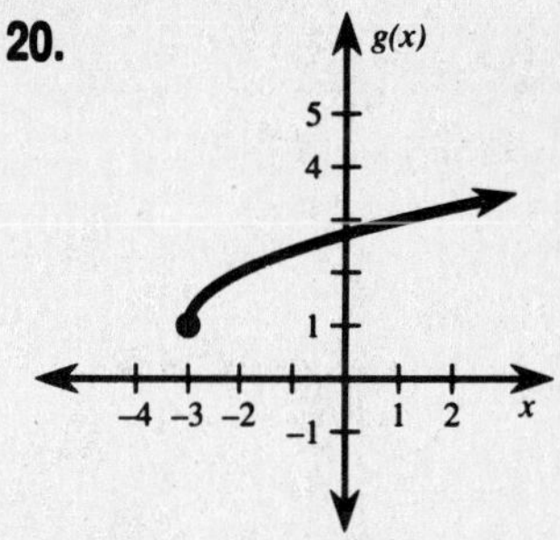

Chapter Test 7-C

1. $-\frac{x^6}{27}$ 2. $\frac{4}{5x^3y^2}$ 3. $\frac{3}{2}$

4. \$1819.40 5. \$487.48 6. \$21,023.65

7. $\frac{1}{32}$ 8. 0.014 9. 5.03 in

10. $\sqrt[4]{11}$ 11. 3.717 12. $\frac{1}{3}$

13. $5\sqrt[4]{2}$ 14. $\frac{1}{7^{1/6}}$ 15. $\frac{2}{3}$

16. $\sqrt[3]{2}$ or $\frac{2}{\sqrt[3]{4}}$

17. 1

18. $x \ge -1$

19. $h(x) \le 2$

20.

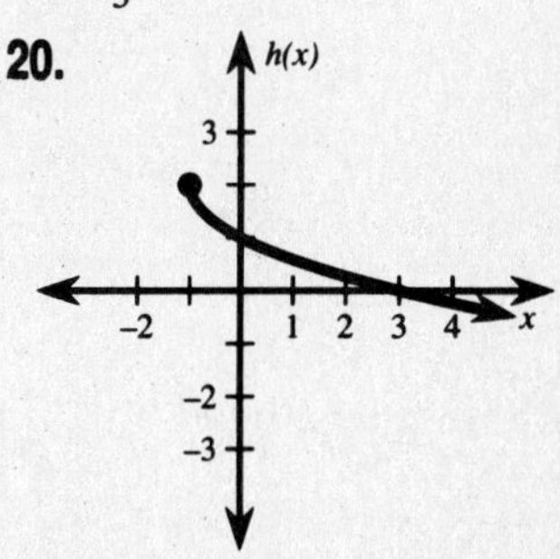

Mid-Chapter Test 8-A

1. 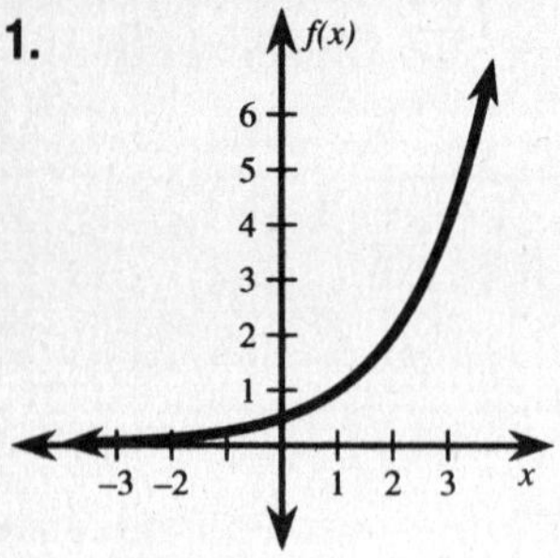

2. 77,494.076
3. \$5624.12
4. $81 = 3^4$
5. $\frac{3}{2}$
6. 1.356
7. $x = 2$
8. $\log_6 5 + \log_6 x - \log_6 y$
9. $\log_{10} \frac{8\sqrt[3]{x}}{y}$ 10. $x = \frac{1}{3}$

Mid-Chapter Test 8-B

1. 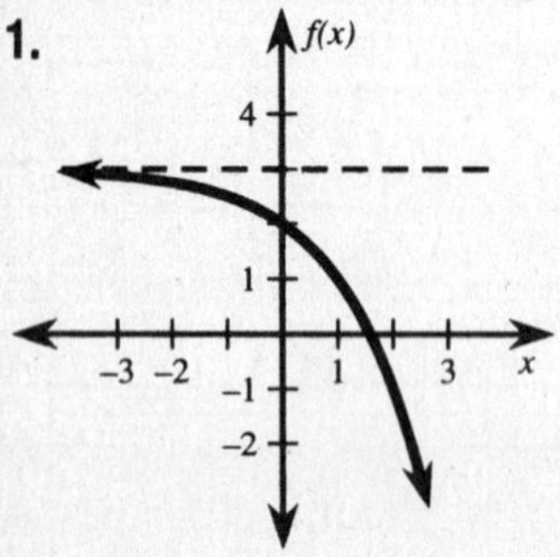

2. 263.817
3. \$7444.17
4. $\log_8 12.126 \approx 1.2$
5. -2
6. 0.845
7. $x = \frac{3}{2}$
8. $\log_5 2 + \frac{1}{2}\log_5 x$
9. $\log_4 \left(\frac{7x^3}{\sqrt{y}}\right)$ 10. $x = \frac{1}{2}$

Chapter Test 8-A

1. 8.807 2. \$5,000,000
3. 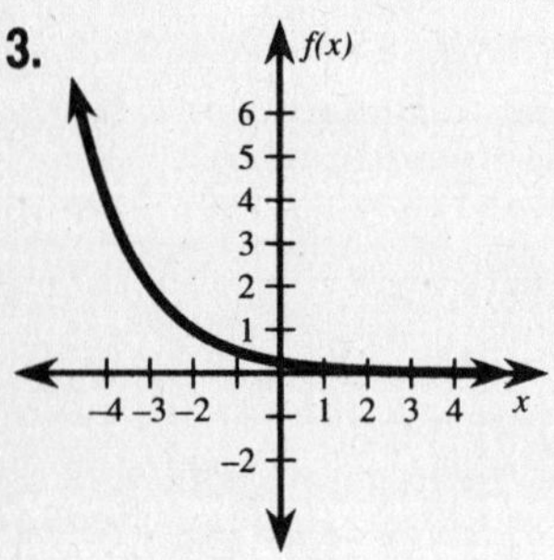

4. $\log_4 \frac{1}{64} = -3$
5. 0.3
6. $\log_4(-4)$
7. 1.362
8. $x = -2$
9. $-2\log_3 x + 3\log_3 y$ 10. $\log_5 \frac{4y^4}{x^3}$
11. $x = 36$
12. $\frac{1}{4e^{4x}}$
13. 3.857
14. Decay
15. 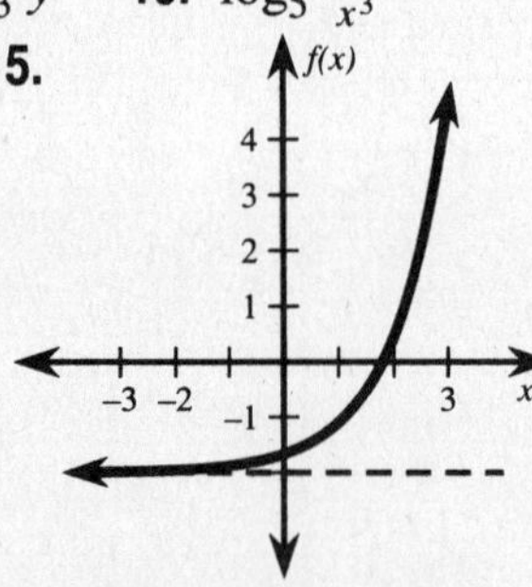

16. $\ln 3 + \ln x + 2\ln y - \ln 19$
17. $\ln\left(\frac{\sqrt[3]{x}}{y^2}\right)$ 18. 10.007 19. 1.654
20. 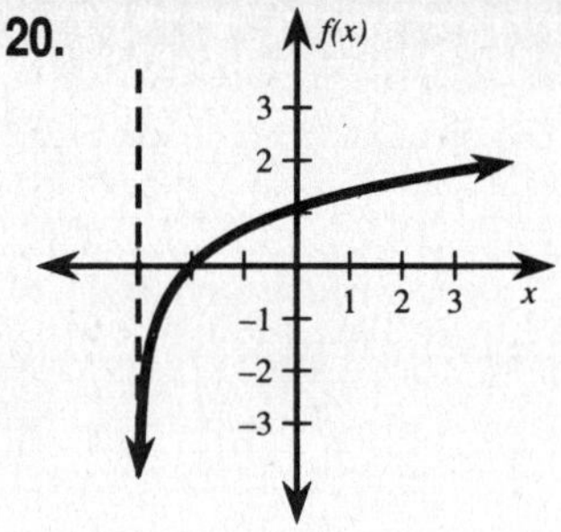

21. $x = 2.513$
22. 11.55 years
23. Asymptotes: $y = 0$, $y = 5$

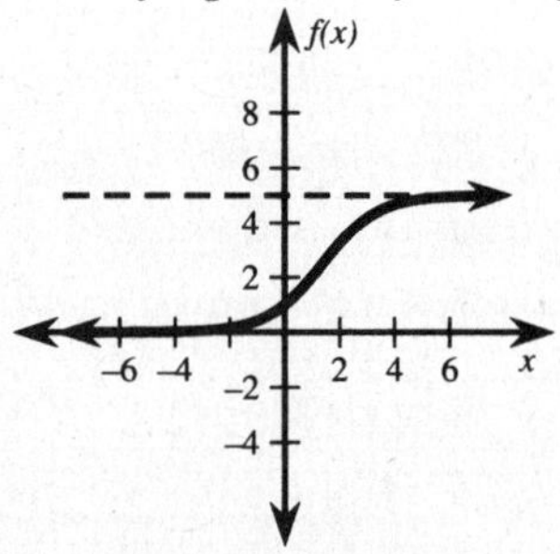

Chapter Test 8-B

1. 4.030 2. \$6,300,000
3. 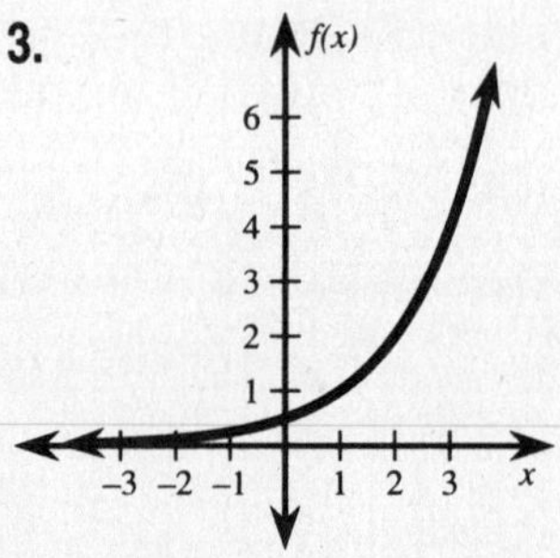

4. $\frac{2}{3} = \log_{27} 9$
5. -2
6. $\log_3(-3)$
7. 1.898
8. $x = -3$
9. $\log_4 3 - \log_4 x + 3\log_4 y$
10. $\log_5 \left(\frac{2y^4}{\sqrt{x}}\right)$
11. $x = 64$
12. $\frac{1}{2e^x}$
13. 12.554
14. Growth
15. 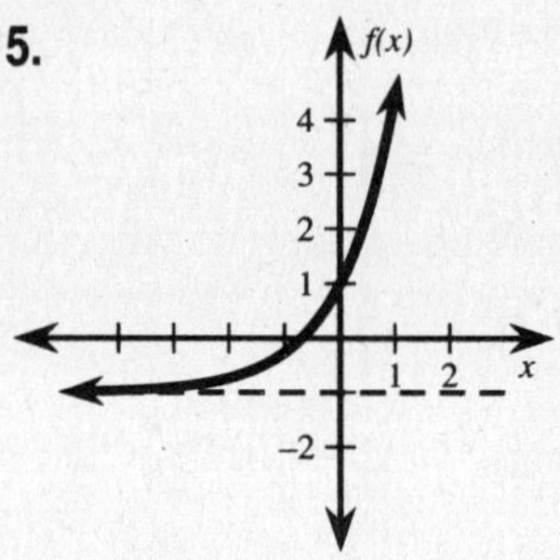

16. $\ln 5 + 2\ln y + 3\ln x - \ln 4$
17. $\ln\sqrt{\frac{x}{y^3}}$
18. 40.309
19. 2.197
20. 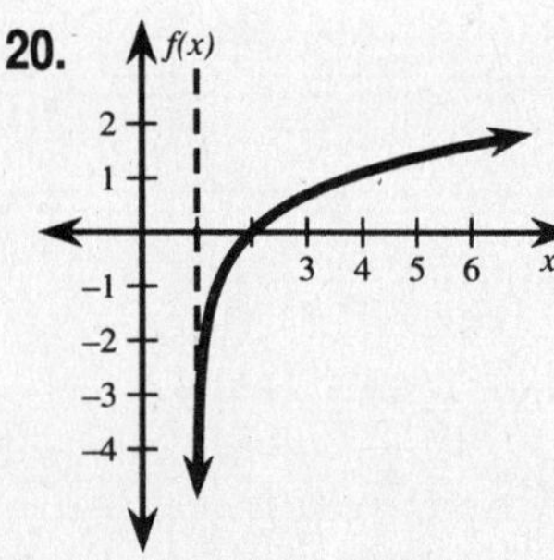

21. 4.868

22. 9.90 years

23. Asymptotes: $y = 2$, $y = 0$

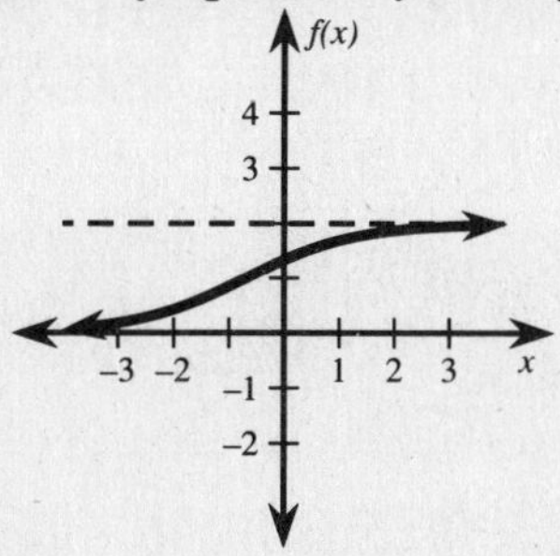

■ Chapter Test 8-C

1. 0.10 **2.** \$103,000

3.

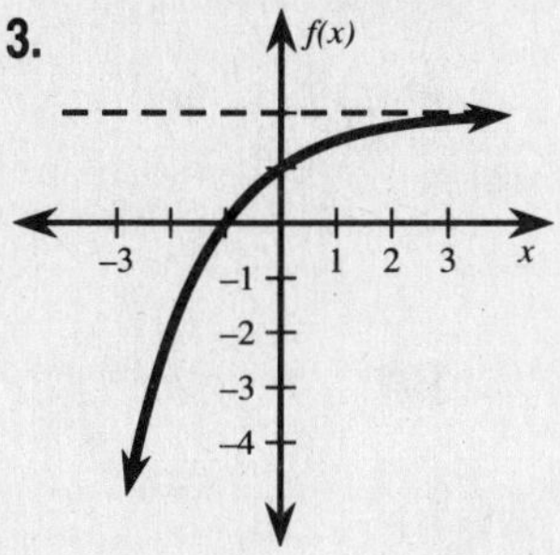

4. $\log_{64} \frac{1}{16} = -\frac{2}{3}$

5. $\sqrt{2}$

6. $\log_4(-6)$

7. 2.681

8. $x = \frac{3}{2}$

9. $1 + \frac{1}{2}\log_5 x - \frac{1}{2}\log_5 y$

10. $\log_3 \frac{2\sqrt{y}}{x^2}$

11. $x = 16$

12. $\frac{125e^{2x}}{2}$

13. 535.492

14. Decay

15.

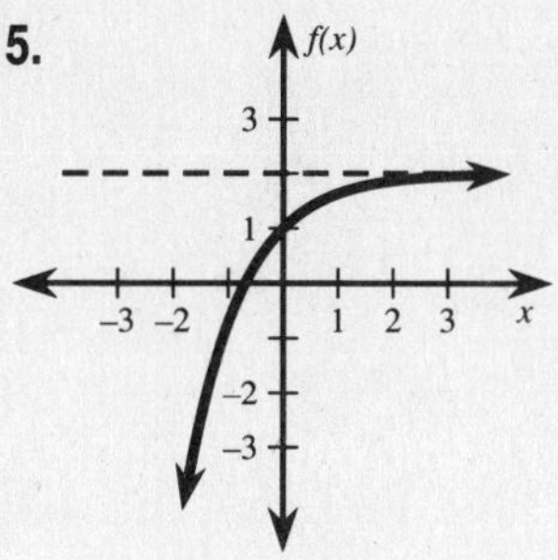

16. $\ln \sqrt[3]{\frac{x^2}{2y}}$ **17.** 2.309 **18.** 2.286

19.

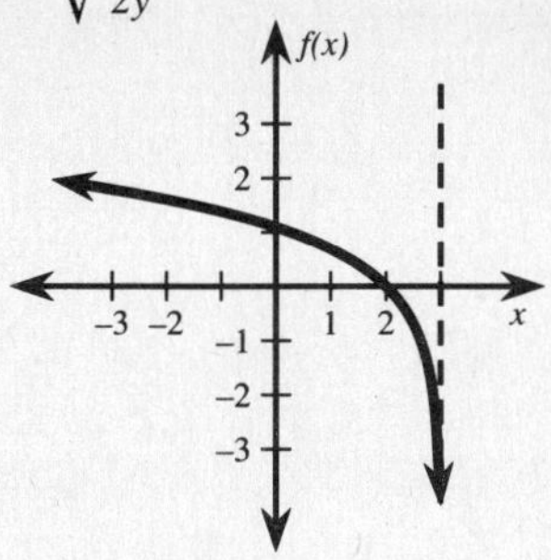

20. 62.712

21. $x = -6.425$

22. 11.64 years

23. Asymptotes: $y = 3$, $y = 0$

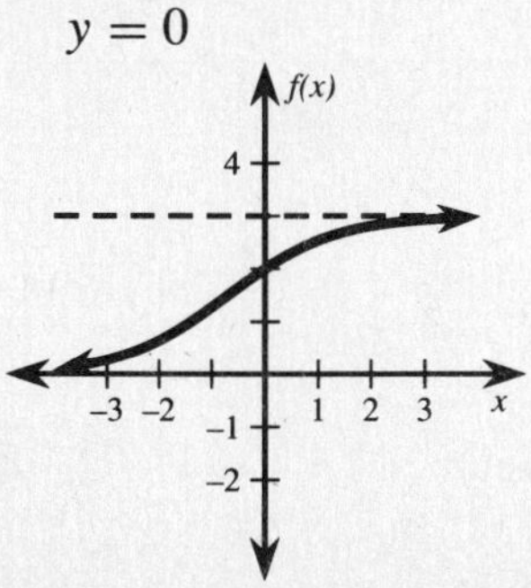

■ Mid-Chapter Test 9-A

1. $7x^3 - x^2 - 2x + 7$

2. $x^3 - x^2 - 18x + 8$

3.

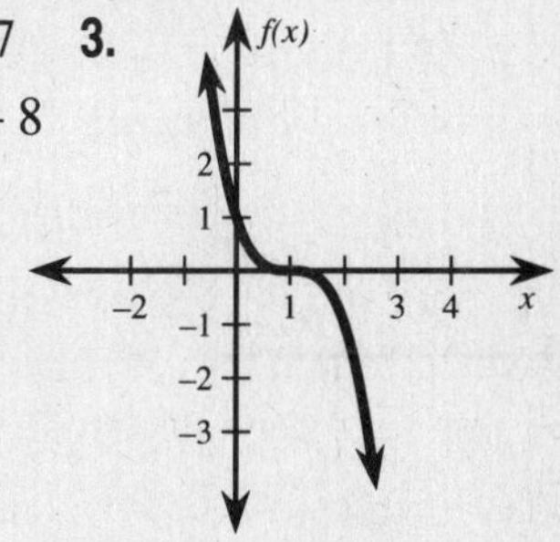

4. $2(4x - 1)(16x^2 + 4x + 1)$ **5.** $-2, \pm 3$

6. $x - 3 + \frac{-2}{x^2+2}$ **7.** $3x + 4 + \frac{19}{x-3}$

8. -4;

3	4	−8	−9	−12	5
		12	12	9	−9
	4	4	3	−3	−4

■ Mid-Chapter Test 9-B

1. $8x^3 + 5x^2 - 3x - 2$

2. $x^3 - x^2 - 7x + 15$

3.

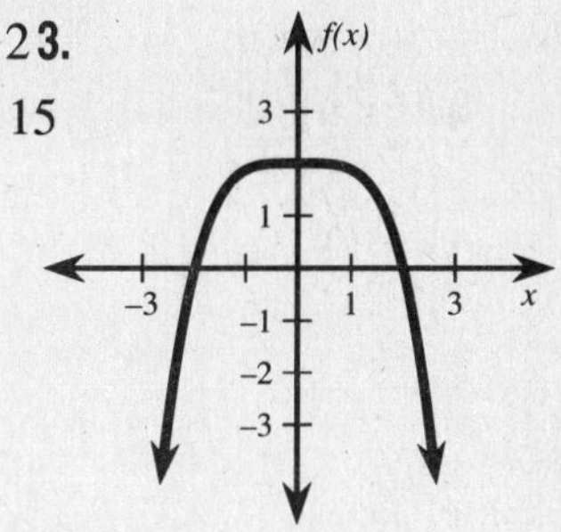

4. $3(3x + 2)(9x^2 - 6x + 4)$ **5.** ± 2

6. $x + 2 + \frac{10}{x^2+4}$ **7.** $5x - 12 + \frac{17}{x+2}$

8. 3;

−3	4	8	−9	12
		−12	12	−9
	4	−4	3	3

■ Chapter Test 9-A

1. $-3x^2 - 5x + 10$ **2.** $6x^3 - 17x^2 - 26x - 8$

3. 2

4. Right: falls
Left: rises

5.

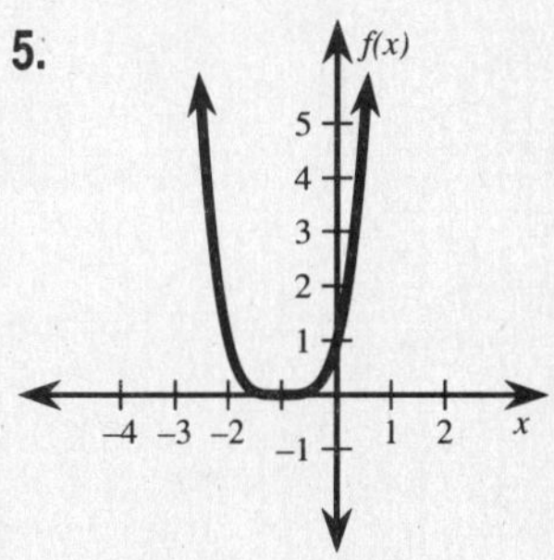

6. $10(x^2 + 4)(x + 2)(x - 2)$

7. $(x - 2)(4x^2 + 3)$ **8.** -2

9. $5x^2 - 13 + \frac{30}{x^2+2}$ **10.** $2x^2 + x - 1 - \frac{2}{x+4}$

11. -9 **12.** $f(x) = x^3 - x^2 - 4x + 4$

13. $(x - 3)(x + 3)(x - 2)$

14. $\frac{1}{2}$, $\pm\sqrt{5}$ **15.** Range: 9; Mean: 47.6
Standard deviation: 2.69

16. $5x^2 - 11x + 4$

■ Chapter Test 9-B

1. $-5x^2 + 9x + 5$ **2.** $12x^3 - 11x^2 - 14x + 4$

3. 3

4. Right: rises
Left: falls

5.

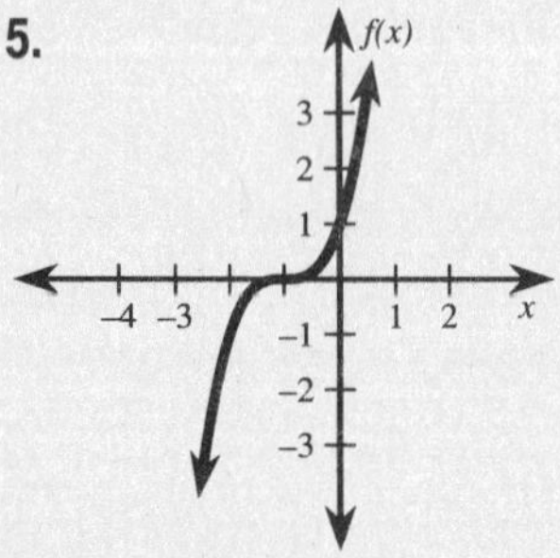

6. $5(x^2 + 3)(x - 1)(x + 1)$

7. $2(x - 2)(x + 2)(x^2 - 3)$

8. 3 **9.** $6x^2 + 2 - \frac{10}{x^2+3}$

10. $2x^2 - 4x + 4 + \frac{4}{x+5}$

11. -6

12. $f(x) = x^3 - 2x^2 - 15x$

13. $(x - 2)(x + 2)(x + 3)$

14. $-\frac{1}{2}$, $\pm\sqrt{3}$ **15.** Range: 10; Mean: 33.8
Standard deviation: 2.96

16. $5x^2 - 13x + 5$

■ Chapter Test 9-C

1. $-11x^2 - 2x + 4$ **2.** $2x^4 - x^3 + 6x^2 - 3x$

3. 4

4. Right: falls
Left: rises

5.

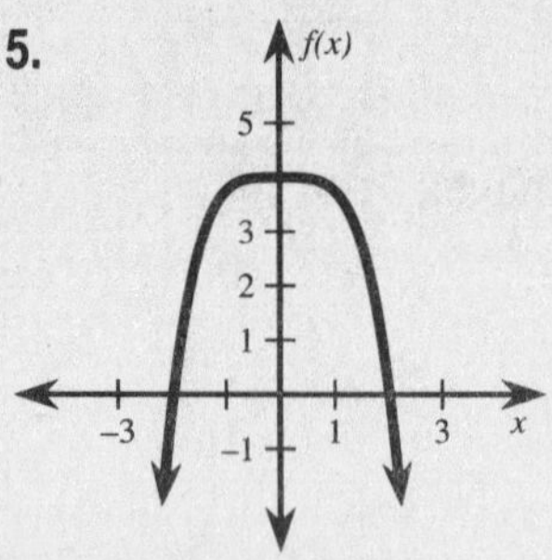

6. $(x - 2)(x + 2)(x - 3)(x + 3)$

7. $(2x - 3)(x^2 + 2)$

8. 1; 2 **9.** $2x^2 + x + 3 + \frac{3x+9}{2x^2-3}$

10. $2x^2 + x - 2 + \frac{7}{x-4}$

11. 10

12. $f(x) = x^3 + \frac{x^2}{6} - \frac{2x}{3} + \frac{1}{6}$

13. $(x - 2)(2x - 1)(2x + 1)$ **14.** $1, -\frac{1}{2} \pm \frac{\sqrt{5}}{2}$

15. Range: 7.7; Mean: 68.82
Standard deviation: 2.34

16. $\frac{7}{2}x^2 + x - \frac{1}{2}$

■ Cumulative Test 7–9

1. $-\frac{1}{64x^6}$ **2.** $\frac{27x^2}{y^3}$ **3.** $x = 7$

4. \$8000 **5.** \$2070.63 **6.** 32

7. 5.241 **8.** 3.63 in. **9.** 3.738

10. 7

11. $4\sqrt[3]{2}$

12. $y = 7$

13. 13

14. 13.75

15. Domain: $x \geq -1$
Range: $g(x) \leq 2$

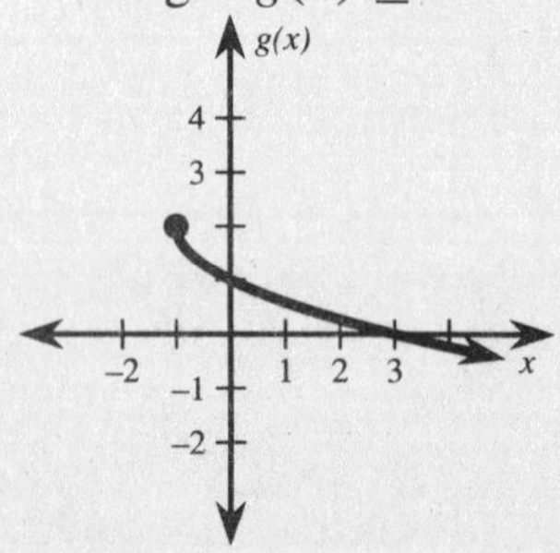

16. $\frac{1}{3}\ln x + \frac{2}{3}\ln y - \ln 3$ **17.** $x = -3$

18. $\log_3 \sqrt{\frac{x^5}{y^2}}$

19. 33.12

20. Growth

21. $x = 144$

22.

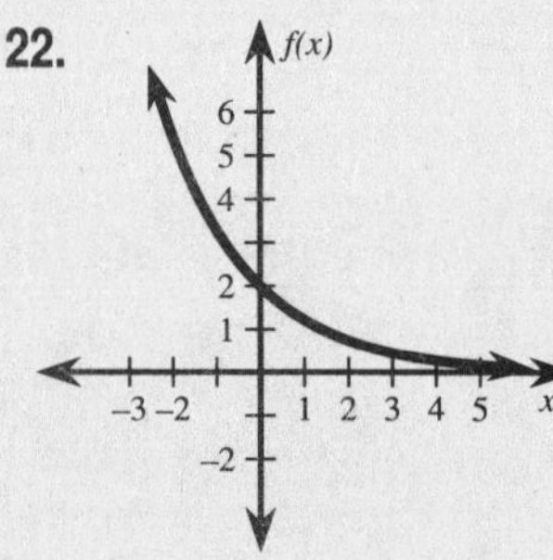

23. 2.775 **24.** 2.467 **25.** 12.60 years

26.

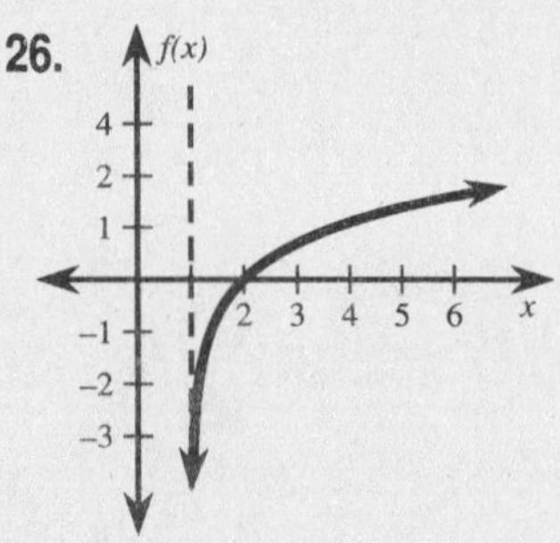

27. Asymptotes:
$y = 0$, $y = 3$

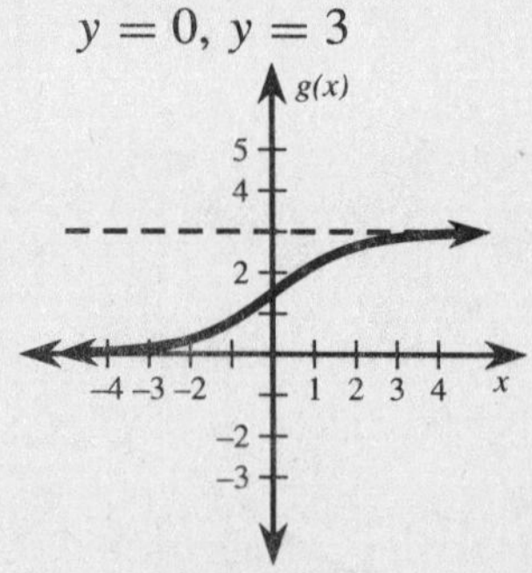

28. $2x^3 - 6x^2 + 13x - 12$

29. 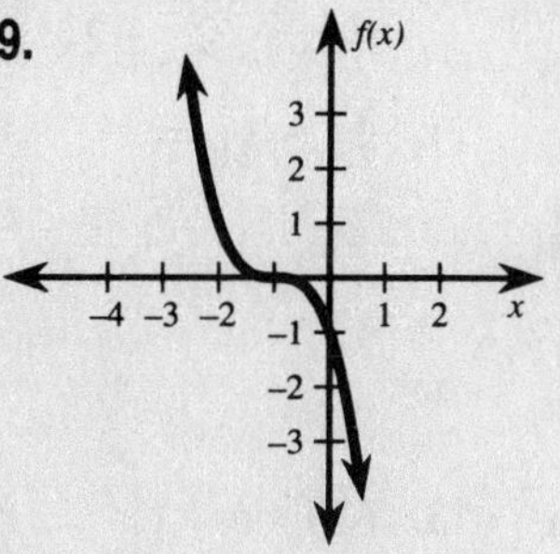

30. 3

31. $3(x-3)(x^2+3x+9)$

32. $(2x-3)(x^2+3)$

33. $-1, 2$

34. $2x^2 + x + \frac{4x}{x^2-4}$

35. -2

36. $x^3 - 2x^2 - 9x + 18$ 37. $x = 3,\ x = \frac{1}{2}$

38. $x = 3,\ x = \pm\sqrt{3}$ 39. $(x-2)(x+3)(x-4)$

40. Range: 12.9; Mean: 33.0
Standard deviation: 3.727

■ Mid-Chapter Test 10-A

1. $x = 0,\ x = 3,\ y = 0$

2. 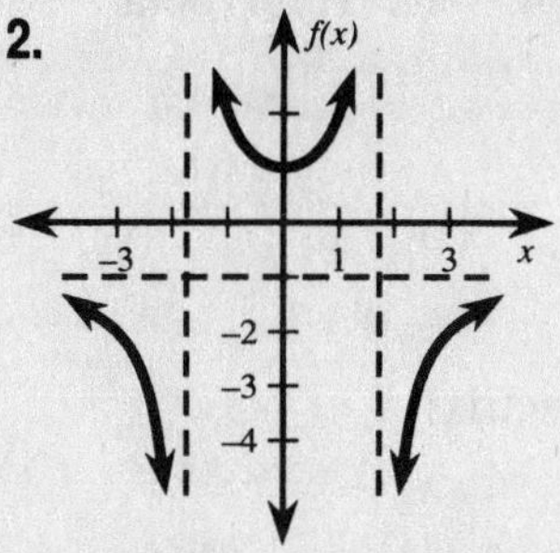

3. $t = \frac{4000}{r}$
or $rt = 4000$;
100 min

4. $z = 25xy$

5. $\frac{x^2-9}{2x}$

6. $\frac{1}{x+2}$

■ Mid-Chapter Test 10-B

1. $x = 0,\ x = 2,\ y = 0$

2. 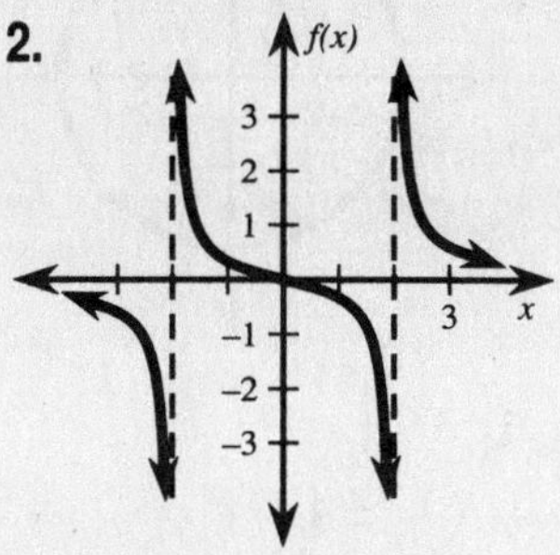

3. $t = \frac{4800}{r}$
or $rt = 4800$;
120 min

4. $z = 7xy$

5. $\frac{3}{x-2}$

6. $2x - 3$

■ Chapter Test 10-A

1. $x = 2,\ x = -2,\ y = 1$

2. 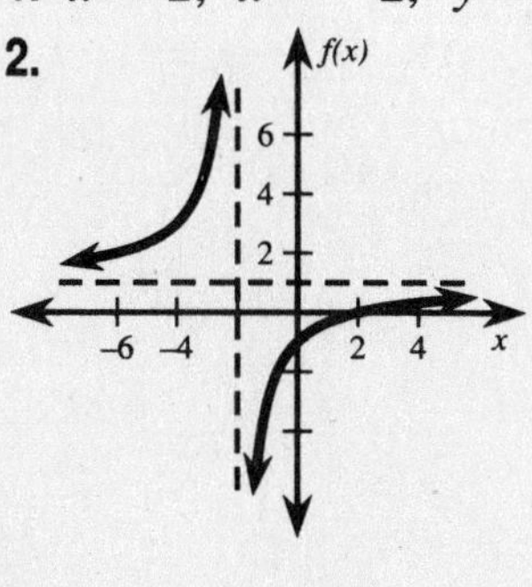

3. $y = \frac{-28}{x}$
or $xy = -28$

4. $z = 0.4xy$

5. $\frac{x-3}{x-1}$

6. $\frac{x(x+2)}{x-5}$

7. $\frac{(x-2)^2}{5x^3}$ 8. $\frac{y}{4x^3}$ 9. $3x^3(x^2-4)$

10. It is not. 11. $x = 8,\ x = -5$

12. $x = -1,\ x = 6$ 13. 2000

14. $\frac{1}{x+4}$ 15. $\frac{5}{3x+4}$ 16. $x = \frac{5}{2}$

17. \$147,725.60 interest 18. \$43,137.60

19. \$222.44 20. \$3,346.40

■ Chapter Test 10-B

1. $x = 1,\ x = -1,\ y = 0$

2. 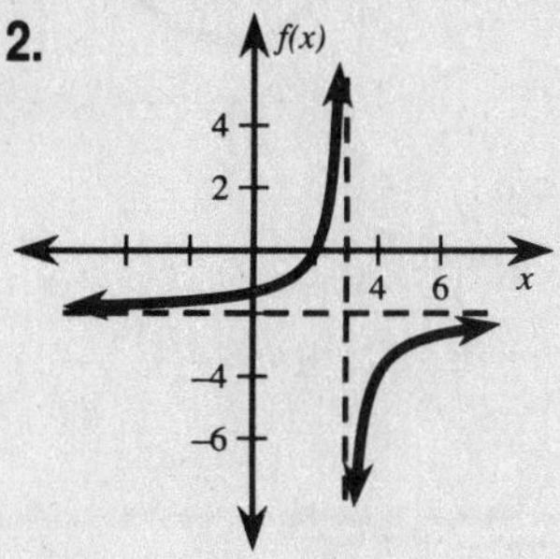

3. $y = -\frac{72}{x}$
or $xy = -72$

4. $z = \frac{1}{200}xy$

5. $\frac{x^2}{x-4}$

6. $\frac{x+1}{(x+3)(2x+3)}$

7. $\frac{1}{(x-2)^2}$ 8. $\frac{2}{x}$ 9. $4x^3(x-1)^2$

10. It is not. 11. $x = -2,\ x = 6$

12. $x = -\frac{1}{3},\ x = \frac{1}{5}$ 13. 4000

14. $\frac{1}{x-5}$ 15. $\frac{2(x+3)}{15}$ 16. $x = -\frac{4}{3}$

17. \$172,610 interest 18. \$41,005.20

19. \$162.53 20. \$5,602.88

■ Chapter Test 10-C

1. $x = 2,\ y = -1$

2. 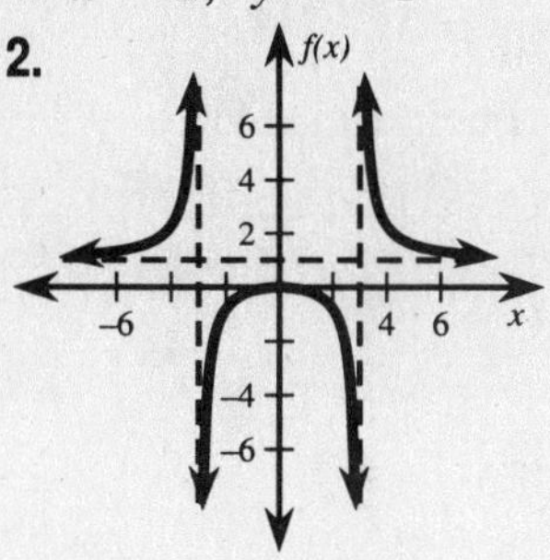

3. $y = \frac{21}{10x}$
or $xy = \frac{21}{10}$

4. $z = \frac{xy}{20}$

5. $\frac{x+3}{x+5}$

6. $\frac{x+1}{3x(x-4)}$

7. $\frac{x}{3(x-2)}$ 8. $\frac{2y^3}{x^5}$ 9. $9x^2(x-3)^2(x+3)$

10. It is a solution. 11. $x = 0,\ x = 2$

12. $x = -1,\ x = -3$ 13. 6000 items

14. $\frac{4}{x+3}$ 15. $-\frac{x+6}{5x+4}$ 16. $x = \frac{9}{14}$

17. 15 years: \$116,030.60
20 years: \$164,261.60

18. \$11,442.60 19. \$251.23 20. \$2,059.04

Mid-Chapter Test 11-A

1. Vertex: (0, 0); Axis: $x = 0$
2. Focus: (0, −4); Directrix: $y = 4$
3.
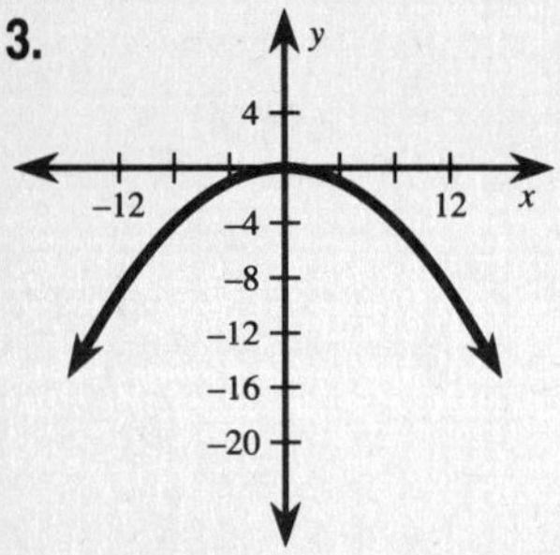

4. $x^2 + y^2 = 29$
5.
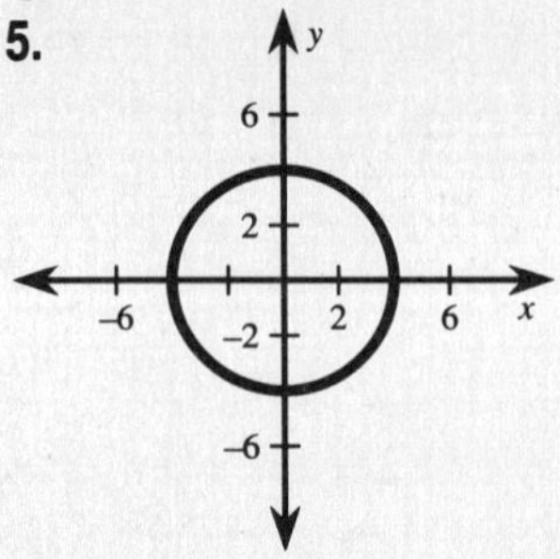

6. Vertices: $(\pm 4,\ 0)$, Foci: $(\pm 2\sqrt{2},\ 0)$
7. $\frac{x^2}{16} + \frac{y^2}{36} = 1,\ e = \frac{\sqrt{5}}{3}$
8. (2, 1); (−1, −2)

Mid-Chapter Test 11-B

1. Vertex: (0, 0); Axis: $y = 0$
2. Focus: (−3, 0); Directrix: $x = 3$
3.
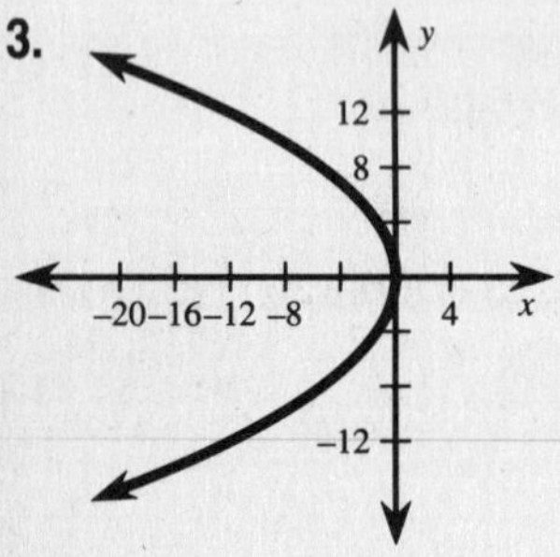

4. $x^2 + y^2 = 13$
5.
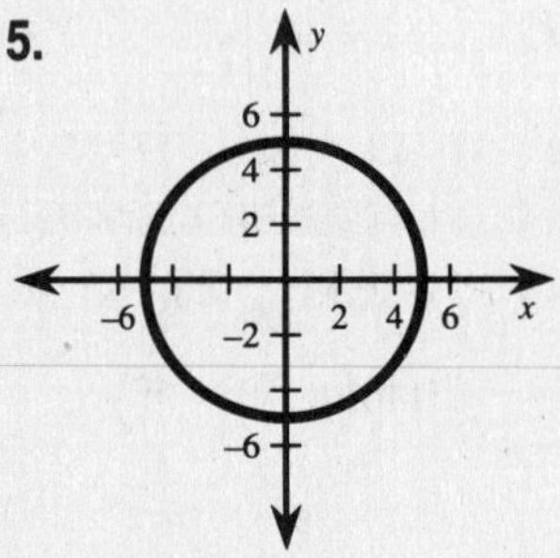

6. Vertices: $(0,\ \pm 2\sqrt{5})$; Foci: $(0,\ \pm 4)$
7. $\frac{x^2}{36} + \frac{y^2}{9} = 1;\ e = \frac{\sqrt{3}}{2}$
8. (2, 4), (−1, 1)

Chapter Test 11-A

1. $x^2 = 20y$
2. Directrix: $x = 1$
   Axis: $y = 0$
3.
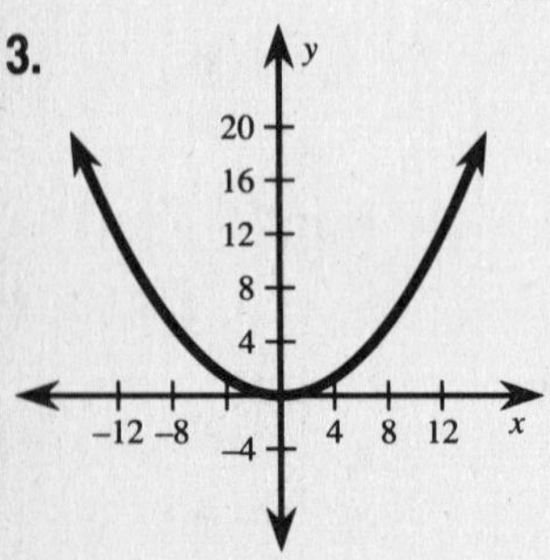

4. $x^2 + y^2 = 49$
5.
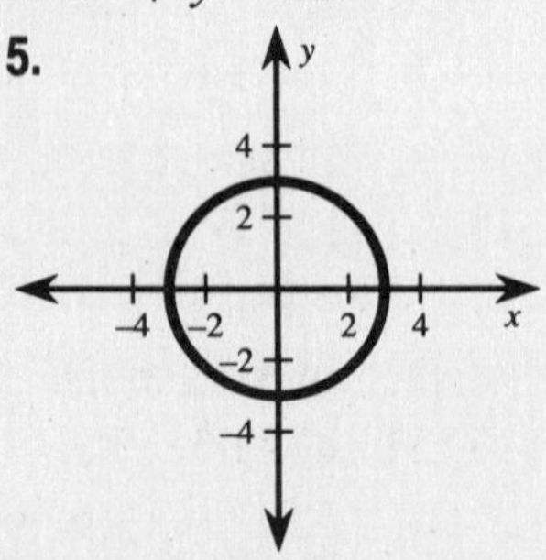

6. (−1, 2)
7. $\frac{x^2}{81} + \frac{y^2}{25} = 1$
8. Vertices: $(\pm 7,\ 0)$; Foci: $(\pm\sqrt{33},\ 0)$; $e = \frac{\sqrt{33}}{7}$
9.
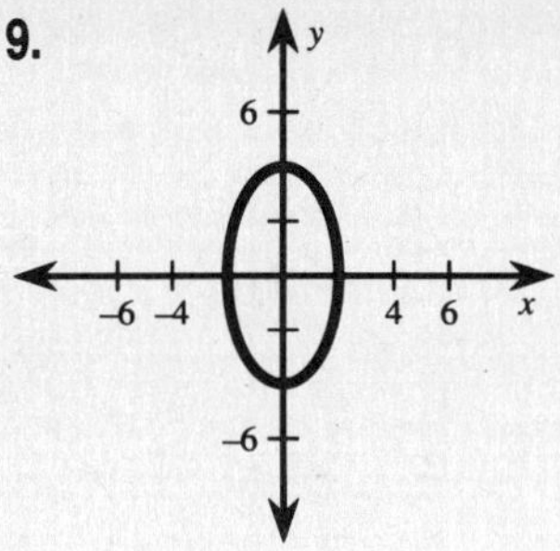

10. Vertices: $(0,\ \pm 2)$
    Foci: $(0,\ \pm 2\sqrt{5})$
11. $\frac{x^2}{25} - \frac{y^2}{24} = 1$
12. Asymptotes: $y = \pm 2x$
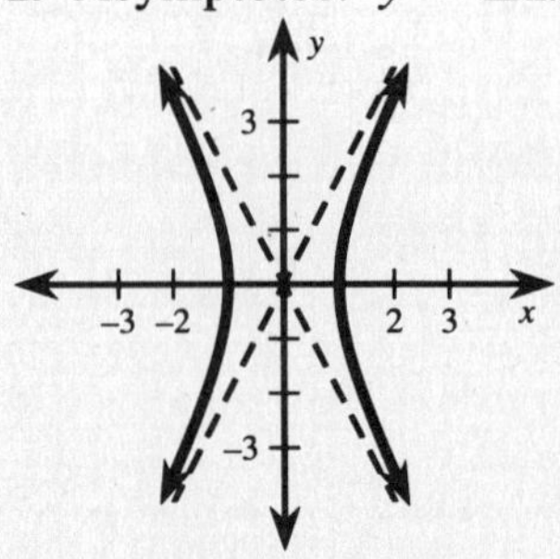

13. Ellipse
    Center: (4, −2)
14. $(y-1)^2 = 8(x+3)$
    (Forms may vary.)
15. $(x-2)^2 + (y+3)^2 = 4$
    Center: (2, −3)
    Radius: 2
16. Hyperbola

Chapter Test 11-B

1. $y^2 = -16x$
2. Axis: $x = 0$
   Directrix: $y = -3$
3.
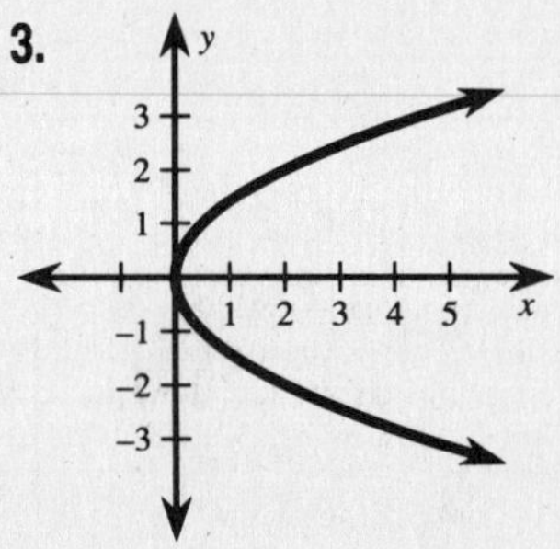

4. $x^2 + y^2 = 81$
5.
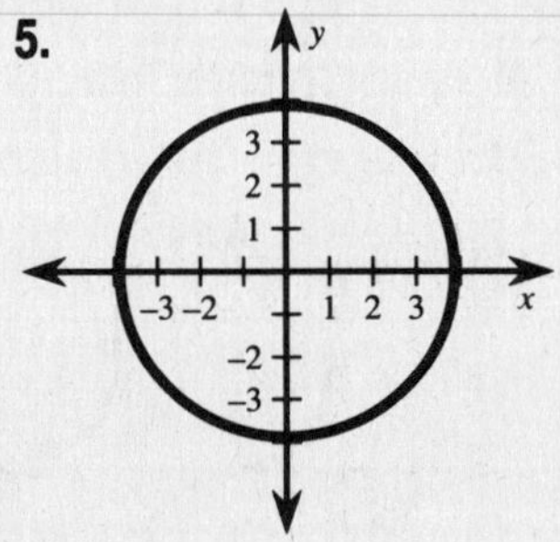

6. (0, 0), (−4, 4)
7. $\frac{x^2}{64} + \frac{y^2}{16} = 1$
8. Vertices: $(0,\ \pm 7)$; Foci: $(0,\ \pm 4)$; $e = \frac{4}{7}$
9.
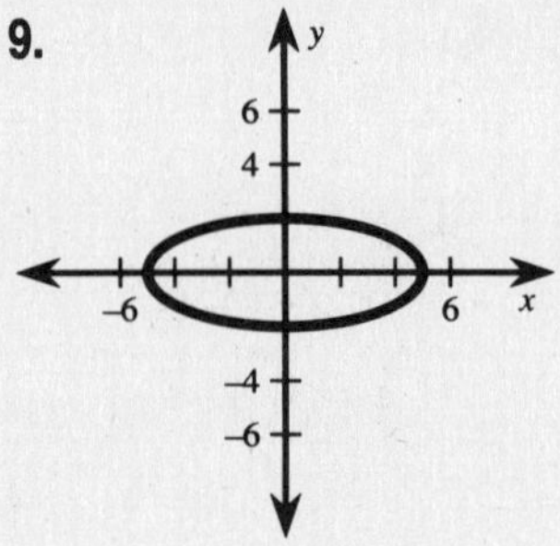

10. Vertices: $(\pm 4,\ 0)$
    Foci: $(\pm 2\sqrt{5},\ 0)$
11. $\frac{x^2}{4} - \frac{y^2}{60} = 1$

12. Asymptotes: $y = \pm\frac{2x}{3}$

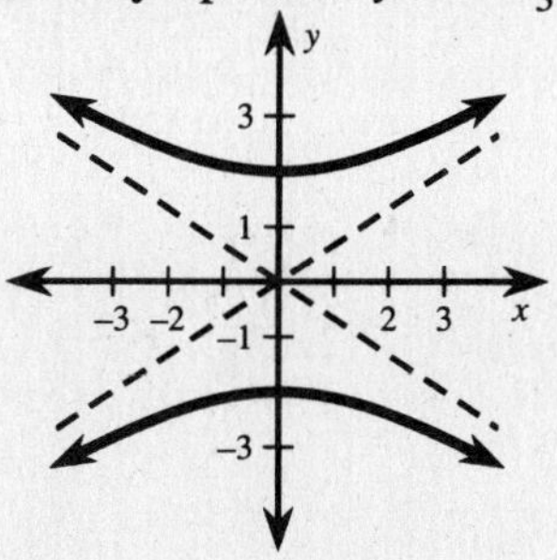

13. Hyperbola
Center: $(-7,\ 8)$

14. $\frac{(x-1)^2}{9} + \frac{(y-2)^2}{4} = 1$

15. $(x-4)^2 + (y+1)^2 = 9$
Center: $(4,\ -1)$
Radius: 3

16. Parabola

■ Chapter Test 11-C

1. $x^2 = 6y$ **2.** Axis: $y = 0$
Focus: $(-\frac{5}{4},\ 0)$

3.

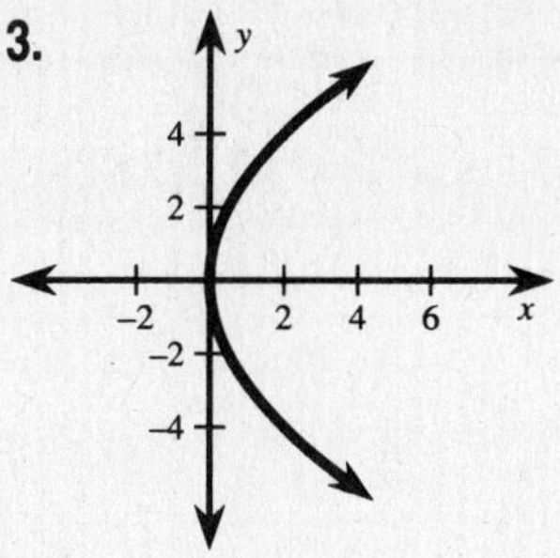

4. $x^2 + y^2 = 37$

5.

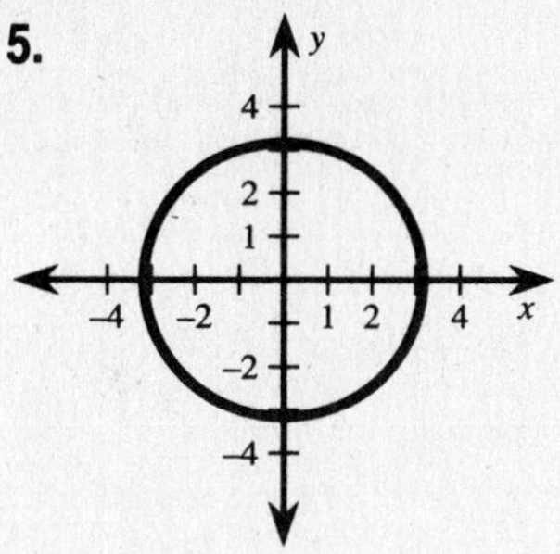

6. $(4,\ -2),\ (-4,\ 2)$ **7.** $\frac{x^2}{25} + \frac{y^2}{9} = 1$

8. Vertices: $(0,\ \pm 5\sqrt{2})$; Foci: $(0,\ \pm 3\sqrt{2})$; $e = \frac{3}{5}$

9.

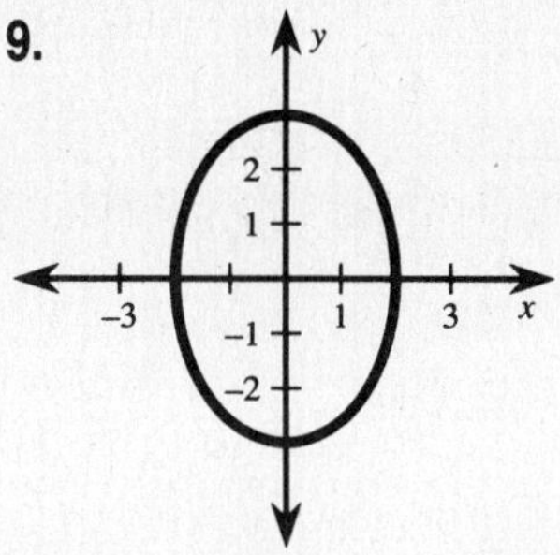

10. Vertices: $(0,\ \pm\sqrt{5})$
Foci: $(0, \pm 4)$

11. $\frac{y^2}{16} - \frac{x^2}{65} = 1$

12. Asymptotes: $y = \pm\frac{1}{2}x$

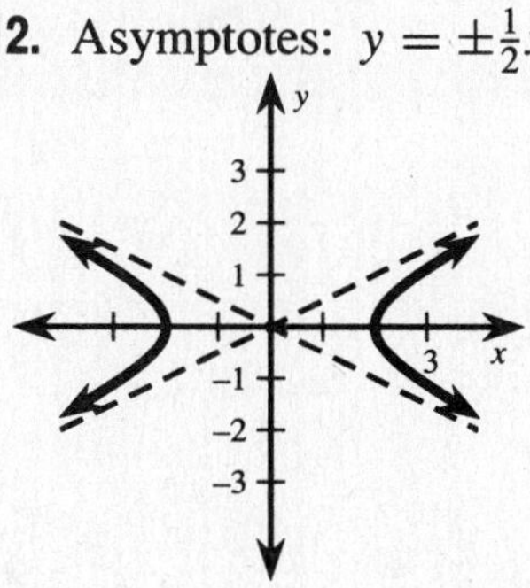

13. Parabola
Vertex: $(-7,\ -3)$

14. $\frac{x^2}{9} - \frac{(y-2)^2}{16} = 1$

15. $(x+3)^2 + (y-3)^2 = 9$
Center: $(-3,\ 3)$
Radius: 3

16. Ellipse

■ Mid-Chapter Test 12-A

1. 2, 4, 7, 11, 16 **2.** 1 + 4 + 9 + 16; 30

3. $a_n = 7n - 2$ **4.** \$1428 **5.** $-\frac{2}{3}$

6. $a_n = 2\left(\frac{3}{2}\right)^{n-1}$ **7.** $a_n = 12\left(\frac{1}{3}\right)^{n-1}$

8. 15.10

■ Mid-Chapter Test 12-B

1. 3, 1, −2, −6, −11 **2.** 2 + 5 + 8 + 11; 26

3. $a_n = 9n - 1$ **4.** \$1112.50 **5.** $-\frac{3}{4}$

6. $a_n = 2\left(\frac{5}{2}\right)^{n-1}$ **7.** $a_n = \frac{3}{2}(2)^{n-1}$ **8.** 14.74

■ Chapter Test 12-A

1. $2, 2, \frac{4}{3}, \frac{2}{3}, \frac{4}{15}$ **2.** 5, 7, 9, 11, 13

3. 0 + 3 + 8 + 15; 26 **4.** $\frac{1}{6}$

5. $a_n = \frac{5}{2}n - \frac{19}{2}$ **6.** 15 **7.** 2300

8. 430 **9.** Neither **10.** −4 **11.** $\frac{5}{16}$

12. $a_n = 24\left(\frac{-3}{4}\right)^{n-1}$ **13.** $\frac{696{,}300}{59{,}049} \approx 11.79$

14. b. $\sum_{n=1}^{\infty} 5\left(\frac{4}{3}\right)^{n-1}$ because $\frac{4}{3} > 1$

15. $\frac{35}{3}$ **16.** $\frac{1}{99}$ **17.** 70

18. $8s^3 - 36s^2t + 54st^2 - 27t^3$ **19.** 120

20. \$16,387.93

■ Chapter Test 12-B

1. $\frac{1}{2}, 1, \frac{9}{8}, 1, \frac{25}{32}$ **2.** 6, 3, 0, −3, −6

3. 3 + 6 + 11 + 18; 38 **4.** $-\frac{1}{3}$

5. $a_n = \frac{2}{3}n + \frac{22}{3}$ **6.** 11 **7.** 3750

8. 395 **9.** Neither **10.** −3 **11.** $-\frac{1}{8}$

12. $a_n = 48\left(-\frac{2}{3}\right)^{n-1}$ **13.** $\frac{58025}{128} \approx 453.32$

14. b. $\sum_{n=1}^{\infty} 6\left(\frac{5}{4}\right)^{n-1}$ because $\frac{5}{4} > 1$

15. 18 **16.** $\frac{1}{49}$ **17.** 126

18. $27s^3 - 54s^2t + 36st^2 - 8t^3$

19. 210 **20.** \$17,308.48

■ Chapter Test 12-C

1. $1, 4, \frac{9}{2}, \frac{8}{3}, \frac{25}{24}$ **2.** 1, 5, 14, 30, 55

3. 0 + 1 + 5 + 23; 29 **4.** $\frac{5}{2}$

5. $a_n = \frac{1}{6}n + \frac{1}{3}$ **6.** 82 **7.** 4065

8. $\frac{825}{4} = 206.25$ **9.** Neither **10.** $\frac{3}{5}$

11. $a_1 = 12,\ a_6 = -\frac{4}{81}$

12. $a_n = 3\left(-\frac{2}{3}\right)^{n-1}$ **13.** $\frac{4947635}{262144} \approx 18.87$

14. b. $\sum_{n=1}^{\infty} (1.01)^{n-1}$ because $1.01 > 1$

15. 7 16. 99 17. 792

18. $A^4 - 8A^3B + 24A^2B^2 - 32AB^3 + 16B^4$

19. 252 20. \$136.10

Cumulative Test 7–12

1. $\frac{4x}{y^2}$ 2. $x = 4$ 3. $\frac{27}{8}$

4. \$3,624.87 5. 29.520

6. 4.976 7. \$18,700 8. $12\sqrt{2}$

9. $\frac{1}{2}\ln x + \frac{1}{4}\ln z - \frac{3}{4}\ln y$

10. 0.659 11. $x = \frac{3}{2}$

12.

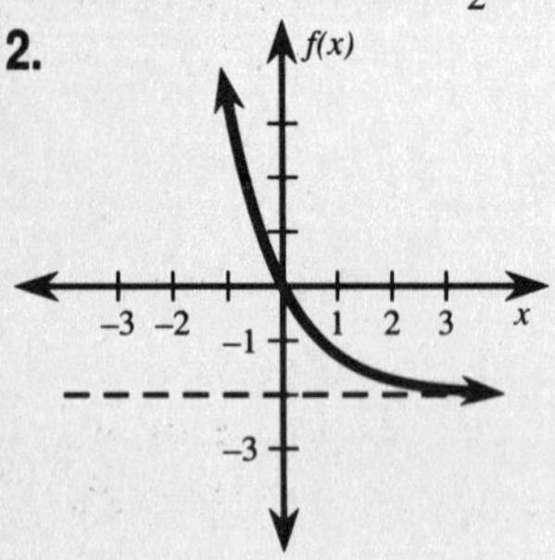

13. 2.324

14.

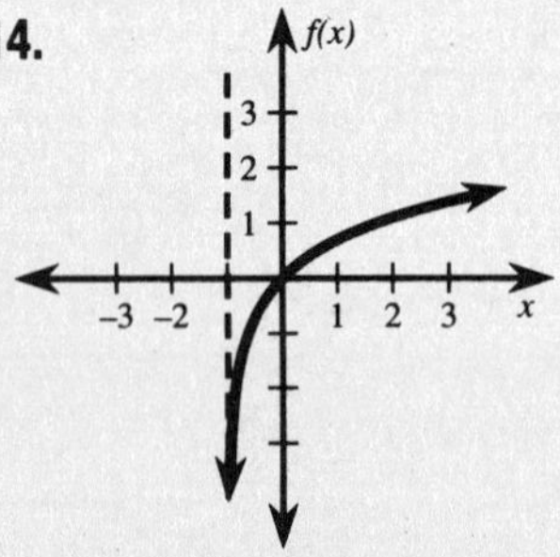

15. $5x^3 + 7x^2 - 5x - 10$ 16. $(x+3)^2(x-3)$

17.

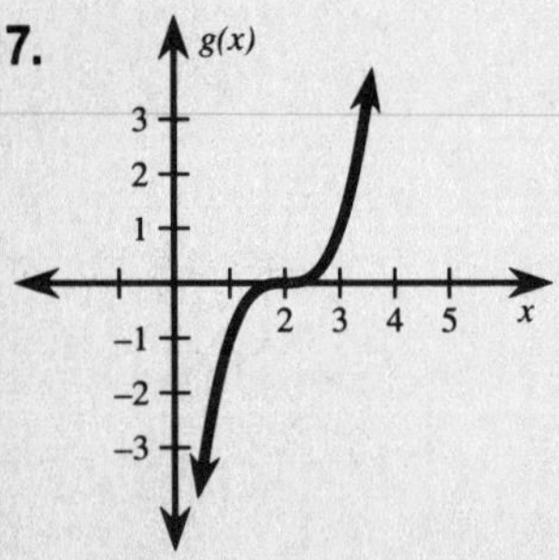

18. $2x^2 - 2x + 1 + \frac{3}{x+6}$

19. $x^3 + 2x^2 - 5x - 6$

20. $x = -2,\ x = \pm\sqrt{5}$

21. Range: 8.60; Mean: 48.77
Standard deviation: 2.43

22. Asymptotes:
$x = -1,\ y = -1$

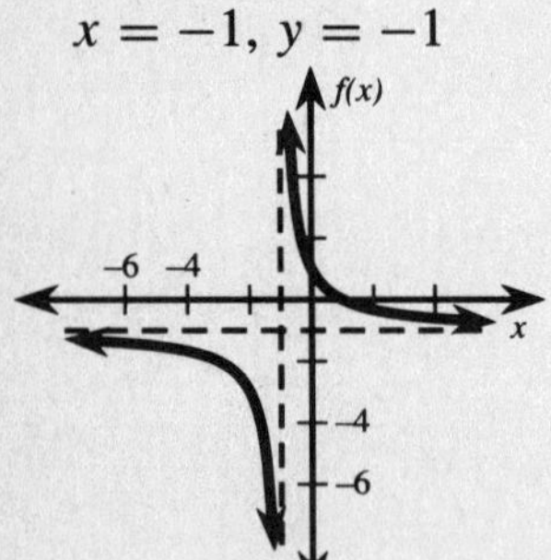

23. $y = \frac{16}{x}$
or $xy = 16$

24. $\frac{1}{x}$

25. $x = 15,\ x = -6$

26. 600

27. $\frac{18-4x}{x+15}$

28. $M = \$66.43$; Total interest, \$391.48

29. $x^2 = 8y$

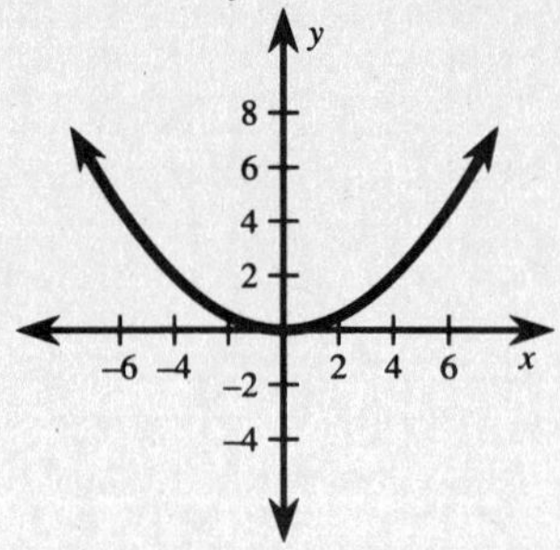

30. $(x-2)^2 + (y+4)^2 = 20$

31. Vertices: $(\pm 6,\ 0)$
Foci: $(\pm 2\sqrt{5},\ 0)$
$e = \frac{\sqrt{5}}{3}$

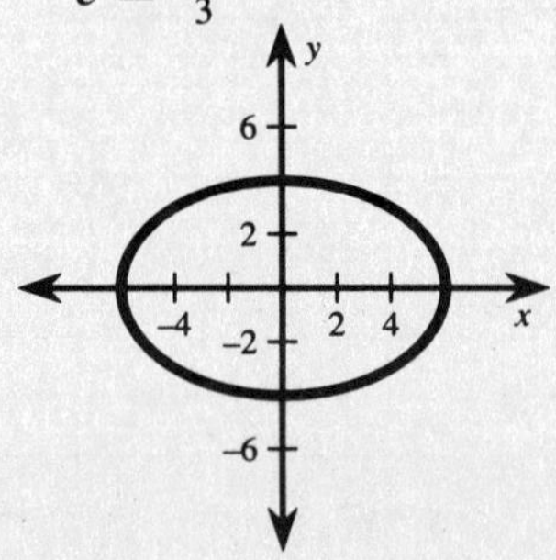

32. Asymptotes: $y = \pm\frac{3}{4}x$
Equation: $\frac{y^2}{9} - \frac{x^2}{16} = 1$

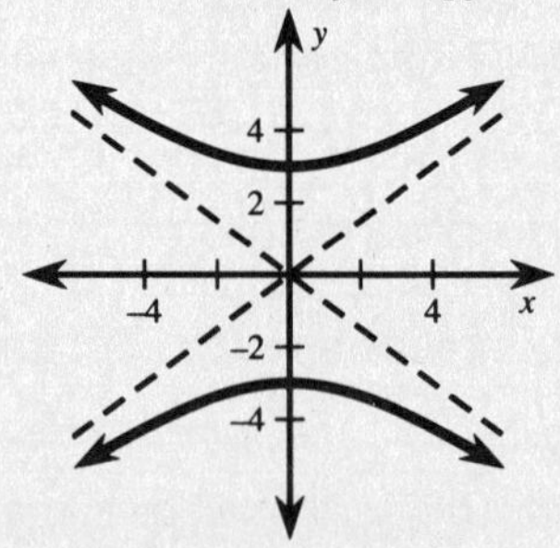

33. Ellipse 34. $1,\ \frac{5}{4},\ \frac{5}{4},\ \frac{17}{16},\ \frac{13}{16}$

35. $a_n = \frac{1}{2}n + \frac{3}{4}$ 36. 1120

37. $a_n = 96\left(-\frac{1}{2}\right)^{n-1},\ a_{12} = \frac{-3}{64}$

38. $\frac{1675674}{46656} \approx 35.92$ 39. 125 40. 252

41. $16x^4 - 96x^3 + 216x^2 - 216x + 81$

Mid-Chapter Test 13-A

1. $\frac{2\sqrt{2}}{3}$ 2. $\sqrt{5}$ 3. 2.559 4. 0.325

5. 40° and 400° 6. $\frac{\pi}{3}$ radians 7. 57°

8. 108° 9. $\frac{1}{2}$ 10. Correct

Mid-Chapter Test 13-B

1. $\frac{\sqrt{5}}{3}$ 2. $\sqrt{6}$ 3. 1.079 4. 1.804

5. −320° and 40° 6. 53° 7. $\frac{\pi}{4}$ radians

8. 67.5° 9. $-\frac{12}{5}$ 10. Not correct

Chapter Test 13-A

1. $\sqrt{2}$ 2. $\frac{1}{2}$ 3. $\frac{1}{\sqrt{3}} = \frac{\sqrt{3}}{3}$ 4. $\frac{2\sqrt{6}}{5}$

5. $\sqrt{5}$ 6. 0.978 7. 2.605 8. 0.208

9. 1.122 10. 2.924 11. 10° and 370° 12. 16°

13. $\frac{\pi}{3}$ radians 14. 54° 15. $\frac{11\pi}{12}$ radians

16. $\frac{4\pi}{15}$ radians 17. 84° 18. $-\frac{1}{\sqrt{2}}$

19. 135° or $\frac{3\pi}{4}$ radians 20. ≈ 95.7

21. ≈ 33.0 22. Area: ≈ 965.4; Side: ≈ 41.6

■ Chapter Test 13-B

1. 1 **2.** $\frac{1}{2}$ **3.** $\frac{2}{\sqrt{3}} = \frac{2\sqrt{3}}{3}$ **4.** $\frac{2\sqrt{2}}{3}$

5. $\frac{\sqrt{13}}{2}$ **6.** 0.306 **7.** 0.358 **8.** 1.022

9. 0.454 **10.** 2.747 **11.** 30° and −330°

12. 64° **13.** $\frac{\pi}{4}$ radians **14.** 144°

15. $\frac{5\pi}{8}$ radians **16.** $\frac{\pi}{5}$ radians **17.** 120°

18. $-\frac{1}{2}$ **19.** −30° or $-\frac{\pi}{6}$ radians

20. ≈ 130.7 **21.** ≈ 31.8

22. Area: ≈ 436.6; Side: ≈ 33.5

■ Chapter Test 13-C

1. 2 **2.** $\frac{1}{\sqrt{2}} = \frac{\sqrt{2}}{2}$ **3.** $\sqrt{3}$

4. $\frac{\sqrt{15}}{4}$ **5.** $\frac{3}{\sqrt{13}} = \frac{3\sqrt{13}}{13}$ **6.** 7.185

7. 0.900 **8.** 0.174 **9.** 1.236

10. 0.934 **11.** $\frac{\pi}{4}$ and $-\frac{7\pi}{4}$ **12.** 45°

13. $\frac{\pi}{18}$ radians **14.** 176° **15.** $\frac{4\pi}{9}$ radians

16. $\frac{\pi}{10}$ radians **17.** 75° **18.** $-\sqrt{5}$

19. −0.304 radians **20.** ≈ 275

21. ≈ 35.0 **22.** Area: ≈ 450; Side: ≈ 32.3

■ Mid-Chapter Test 14-A

1. Amplitude: $\frac{3}{2}$
Period: π

2.

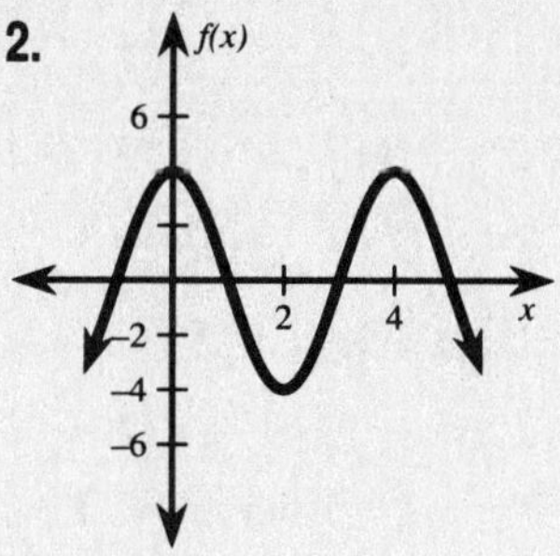

3. $y = 6\cos\frac{4x}{3}$

4. $y = -2 + 7\sin 2x$

5.

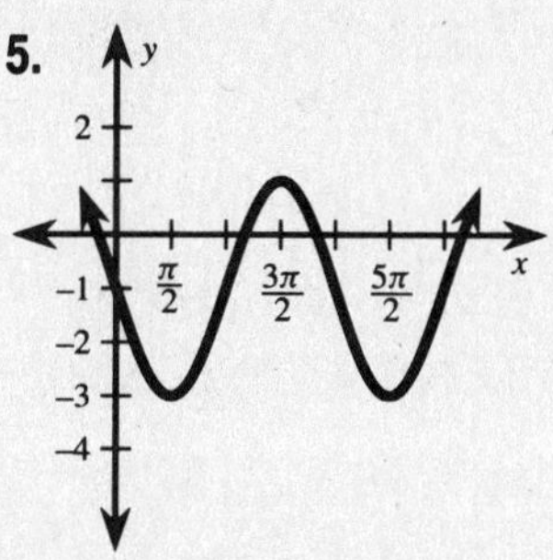

6. $\csc x$

7. $\cos^4 x - \sin^4 x$
$= (\cos^2 x + \sin^2 x)(\cos^2 x - \sin^2 x)$
$= 1(\cos^2 x - \sin^2 x)$

■ Mid-Chapter Test 14-B

1. Amplitude: 5
Period: 4

2.

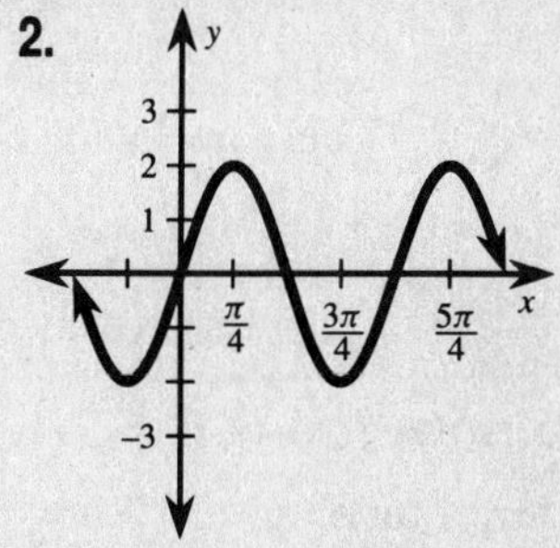

3. $y = 3\sin\frac{2\pi x}{3}$

4. $y = -2\cos 3(x+2)$

5.

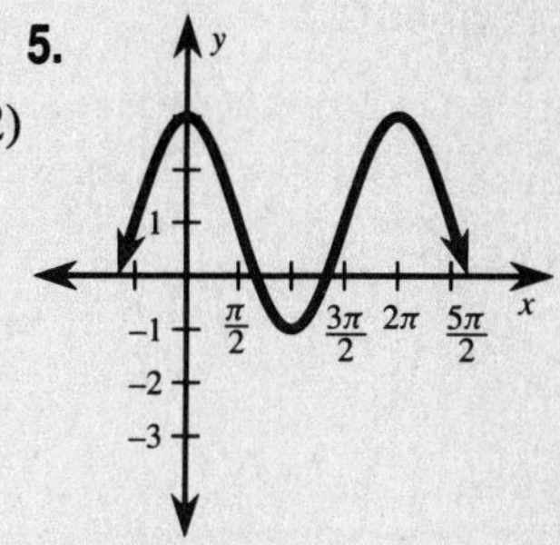

6. $-\sin x$ **7.** $\tan^2 x + \sec^2 x$
$= (\sec^2 x - 1) + \sec^2 x$
$= 2\sec^2 x - 1$

■ Chapter Test 14-A

1. Amplitude: 3
Period: 1

2.

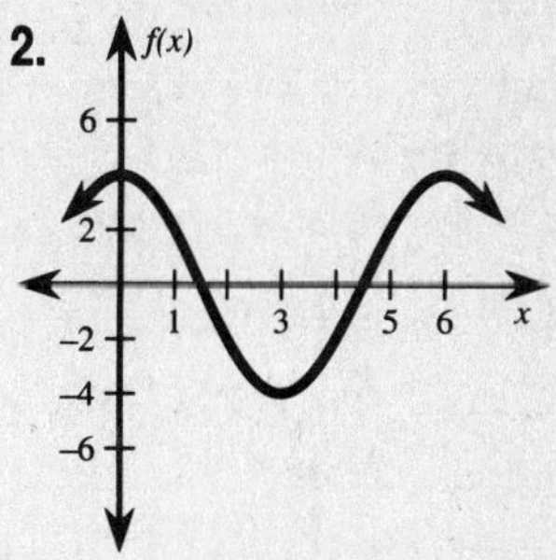

3. $y = 0.6\cos 4x$

4. 0, π
(There are other correct values.)

5. a. $y = \frac{1}{2}\sin 4x$

6. $y = 2 + 6\sin\frac{\pi x}{3}$

7.

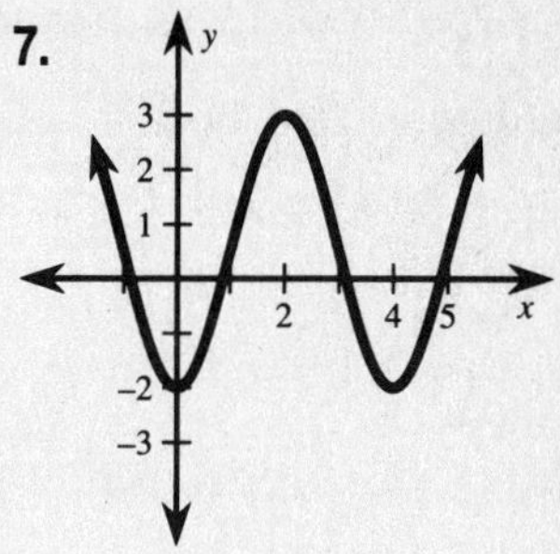

8. 1 **9.** $\cot x$

10. $$\frac{\sin x}{1+\cos x} = \frac{\sin x(1-\cos x)}{(1+\cos x)(1-\cos x)} = \frac{\sin x}{1-\cos^2 x}(1-\cos x) = \frac{(1-\cos x)\sin x}{\sin^2 x} = \frac{1-\cos x}{\sin x}$$

11. $\frac{\pi}{6}, \frac{5\pi}{6}, \frac{7\pi}{6}, \frac{11\pi}{6}$ 12. $\sin 70^\circ$

13. $(0, 0), (\pi, 0), (2\pi, 0), \left(\frac{\pi}{6}, \frac{1}{2}\right), \left(\frac{5\pi}{6}, \frac{1}{2}\right)$

14. $\frac{\sqrt{6}-\sqrt{2}}{4}$ (There are other correct forms.)

15. $\frac{\pi}{6}, \frac{\pi}{2}, \frac{5\pi}{6}, \frac{3\pi}{2}$

16. $\frac{\sqrt{2+\sqrt{3}}}{2}$ (There are other correct forms.)

■ Chapter Test 14-B

1. Amplitude: 3
   Period: 2

2.

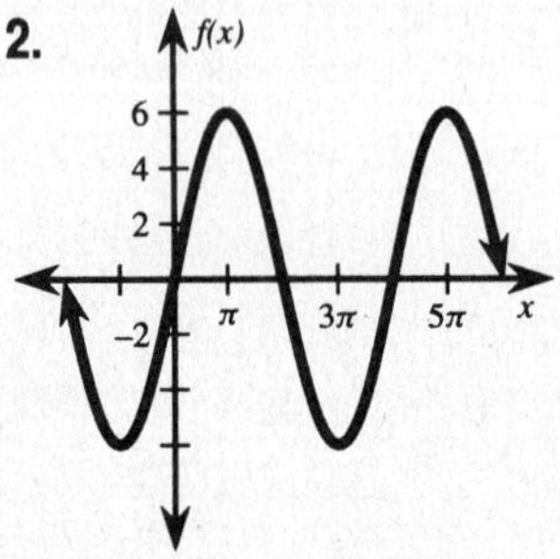

3. $y = \frac{2}{3}\sin\left(\frac{\pi x}{6}\right)$

4. $\frac{3\pi}{4}, \frac{7\pi}{4}$
   (There are other correct values.)

5. b. $y = \frac{1}{2}\cos 4x$

6. $y = -3 - 8\cos 6\pi x$

7.

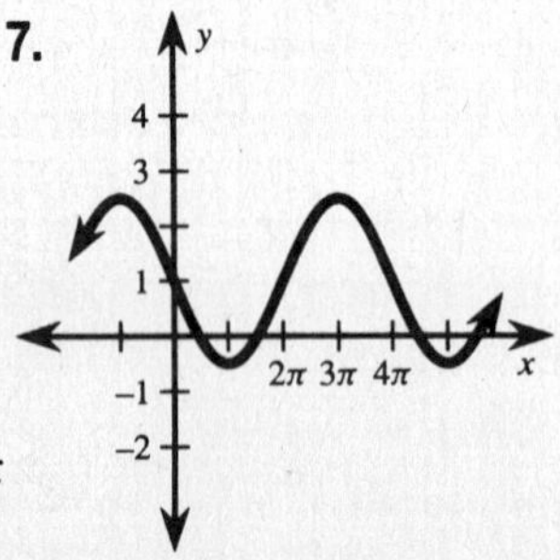

8. $\tan x$ 9. $\cos^2 x$

10. $\tan^2 x \csc^2 x - \tan^2 x = \sec^2 x - \tan^2 x = 1$

11. $\frac{\pi}{3}, \frac{2\pi}{3}, \frac{4\pi}{3}, \frac{5\pi}{3}$ 12. $\cos 80^\circ$

13. $\left(\frac{\pi}{2}, 0\right), \left(\frac{3\pi}{2}, 0\right), \left(\frac{\pi}{3}, \frac{1}{2}\right), \left(\frac{5\pi}{3}, \frac{1}{2}\right)$

14. $\frac{\sqrt{6}+\sqrt{2}}{4}$ (There are other correct forms.)

15. $\frac{\pi}{3}, \pi, \frac{5\pi}{3}$

16. $\frac{\sqrt{2-\sqrt{2}}}{2}$ (There are other correct forms.)

■ Chapter Test 14-C

1. Amplitude: 2
   Period: $\frac{\pi}{3}$

2.

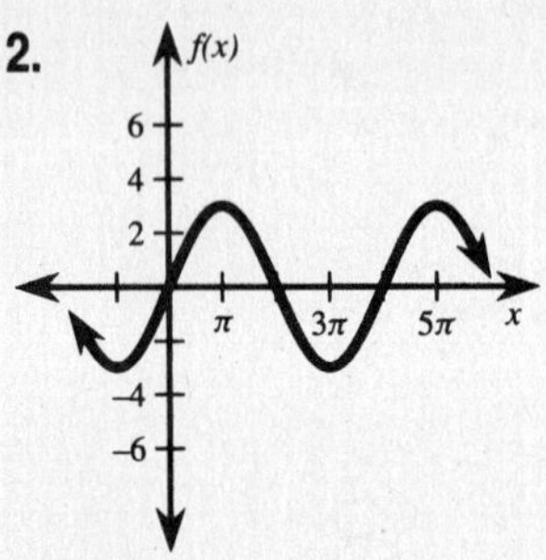

3. $y = \frac{1}{6}\cos 6x$

4. $\frac{\pi}{4}, \frac{3\pi}{4}$

5. b. $y = 4\sin\frac{\pi x}{2}$

6. $y = -\frac{1}{5}\cos\frac{3(x+2)}{8}$

7.

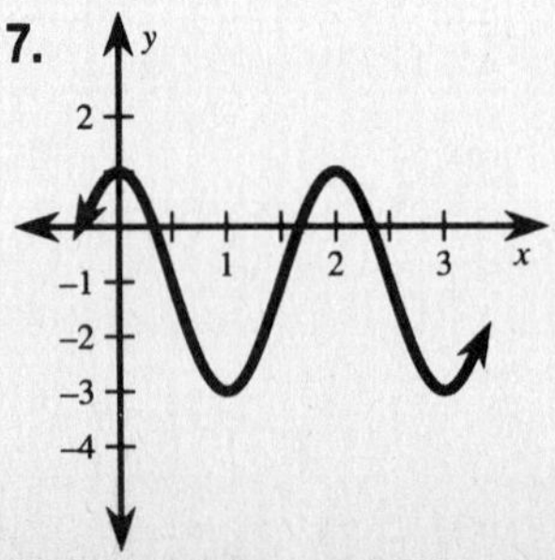

8. $\tan^2 x$ 9. 1

10. $\cos^2 x - \cos^4 x = \cos^2 x(1 - \cos^2 x)$
$$= \cos^2 x \sin^2 x$$
$$= (1 - \sin^2 x)\sin^2 x$$
$$= \sin^2 x - \sin^4 x$$

11. $\frac{\pi}{2}, \frac{7\pi}{6}, \frac{11\pi}{6}$ 12. $\cos\frac{7\pi}{12}$

13. $\left(\frac{\pi}{3}, 1\right), \left(\frac{5\pi}{3}, 1\right)$

14. $2 - \sqrt{3}$ (There are other correct forms.)

15. $0, \frac{\pi}{4}, \frac{3\pi}{4}, \pi, \frac{5\pi}{4}, \frac{7\pi}{4}$

16. $\frac{\sqrt{2-\sqrt{2}}}{2}$ (There are other correct forms.)

■ Mid-Chapter Test 15-A

1. 35 2. 455 3. 2160 4. $\frac{14}{285}$

5. $\frac{\pi}{4}$ 6. $\frac{2}{17}$

■ Mid-Chapter Test 15-B

1. 56 2. 165 3. 1050 4. $\frac{11}{57}$

5. $\frac{4-\pi}{4}$ 6. $\frac{11}{1105}$

■ Chapter Test 15-A

1. 1365 2. $5! = 120$ 3. 4512

4. 90 5. $\frac{40}{52} = \frac{10}{13}$ 6. 32

7. $6! = 720$; $\frac{1}{3}$ 8. $\frac{\pi}{8}$ 9. 0.84

10. 0.28 11. 0.9 12. 0.48 13. $\frac{1}{32}$

14. $\frac{49}{289} \approx 0.17$ 15. \$13.37 16. \$3.50

■ Chapter Test 15-B

1. 1820 2. $6! = 720$ 3. 288 4. 720

5. $\frac{9}{13}$ 6. 5 ways 7. $5! = 120$; $\frac{2}{5}$

8. $\frac{8-\pi}{8}$ 9. 0.76 10. 0.53 11. 0.58

12. 0.224 13. $\frac{31}{32}$ 14. $\frac{100}{289} \approx 0.35$

15. \$18.26 16. \$4.50

■ Chapter Test 15-C

1. 22,100 2. $\frac{5!}{2} = 60$ 3. 704 4. 180

5. $\frac{36}{52} = \frac{9}{13}$ 6. 6; $\frac{6}{36} = \frac{1}{6}$ 7. $8! = 40{,}320$; $\frac{1}{4}$

8. $\frac{2}{\pi}$ 9. 0.11 10. 0.13

11. 0.60 12. $\frac{1}{2}$ 13. $\frac{5}{32}$

14. $\frac{140}{289} \approx 0.48$ 15. 12.55 16. \$5.50

■ Cumulative Test 13–15

1. $\sqrt{2}$ **2.** $\frac{\sqrt{3}}{2}$ **3.** $\frac{\sqrt{3}}{2}$ **4.** $\frac{4}{3}$

5. $\frac{12}{13}$ **6.** 1.942 **7.** 19.081 **8.** 0.809

9. -2.130 **10.** 0.105 **11.** 24°

12. $\frac{\pi}{8}$ radians **13.** 26° **14.** $\frac{2\pi}{5}$

15. 480° **16.** $-\frac{\sqrt{53}}{2}$ **17.** 150°

18. -0.469 radians **19.** 395.56

20. ≈ 69.15 **21.** Area: ≈ 1343.6; $c \approx 44.5$

22. Amplitude: 8.14; Period: 10π

23. $y = 7 \sin 3x$

24.

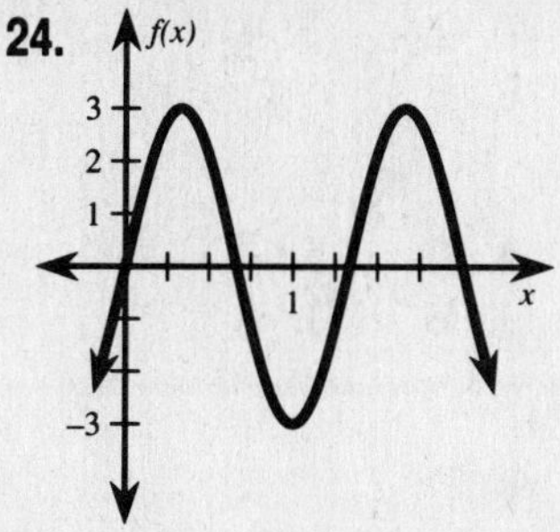

25. $y = \frac{2}{3}\cos(0.3x - 0.6)$

26. $y = \frac{1}{2}\sin 4x$

27.

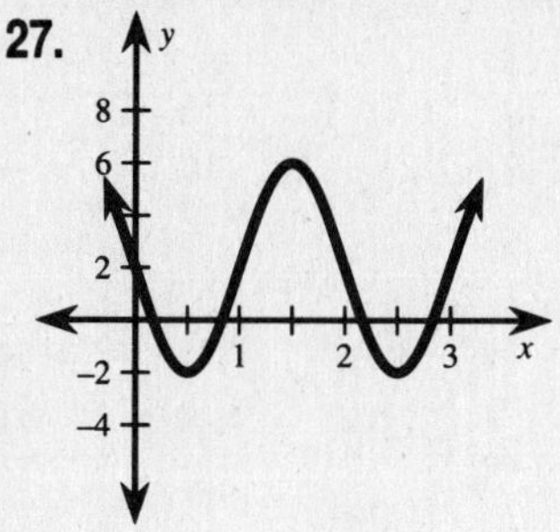

28. $-\sec x$

29. $\cos x$

30. $\cos^4 x - \sin^4 x$

$$= (\cos^2 x - \sin^2 x)(\cos^2 x + \sin^2 x)$$

$$= \cos^2 x - \sin^2 x$$

31. $\frac{\pi}{3}, \frac{2\pi}{3}, \frac{4\pi}{3}, \frac{5\pi}{3}$

32. $\cos 55°$ **33.** $\frac{1+\sqrt{3}}{\sqrt{3}-1}$

34. $\left(\frac{\pi}{6}, \frac{1}{2}\right), \left(\frac{5\pi}{6}, \frac{1}{2}\right), \left(\frac{7\pi}{6}, \frac{1}{2}\right), \left(\frac{11\pi}{6}, \frac{1}{2}\right)$

35. $\frac{\sqrt{6}-\sqrt{2}}{4}$ **36.** $\frac{\pi}{6}, \frac{5\pi}{6}, \frac{3\pi}{2}$

37. $\frac{7!}{2} = 2520$ **38.** 680 **39.** 8450

40. $2^{10} = 1024$ **41.** $\binom{10}{5} = 252$

42. $7! = 5040$; $\frac{2}{7}$ **43.** 0.543 **44.** 0.82

45. 0.95 **46.** 0.45

47. \$16.80 **48.** \$15

Answers to Practice for College Entrance Tests

Chapter 1, Form A

1. C **2.** D **3.** E **4.** D **5.** C
6. A **7.** D

Chapter 1, Form B

1. A **2.** C **3.** A **4.** B **5.** D
6. D **7.** C **8.** A

Chapter 2, Form A

1. C **2.** A **3.** A **4.** A **5.** C
6. A

Chapter 2, Form B

1. $\frac{2}{3}$ **2.** 1 **3.** $\frac{7}{6}$ **4.** 16
5. 4 **6.** $\frac{7}{3}$
7. Any value between but not including 1 and 9 **8.** 34

Chapter 3, Form A

1. D **2.** E **3.** B **4.** E **5.** B
6. A

Chapter 3, Form B

1. A **2.** B **3.** C **4.** C **5.** B
6. D **7.** A **8.** C

Chapter 4, Form A

1. B **2.** B **3.** B **4.** E **5.** C
6. A **7.** D **8.** B **9.** C

Chapter 4, Form B

1. 4 **2.** 91 **3.** 2.37 **4.** 8
5. $\frac{15}{2}$ or 7.5
6. 1 **7.** 7 **8.** $\frac{5}{2}$ or 2.5
9. Any number between and including 10 and 12
10. 3.2

Chapter 5, Form A

1. E **2.** C **3.** D **4.** B **5.** D
6. B **7.** E **8.** A **9.** C

Chapter 5, Form B

1. C **2.** D **3.** A **4.** B **5.** D
6. A **7.** B

Chapter 6, Form A

1. C **2.** E **3.** B **4.** B **5.** D
6. A **7.** E **8.** C

Chapter 6, Form B

1. Any value between, but not including 1 and 5
2. 3 or 4 **3.** 4
4. Any number such that $x \geq 5$
5. 7 **6.** $\frac{3}{28}$ **7.** 19 **8.** 5
9. $\frac{3}{2}$ or 1.5 **10.** $\frac{3}{2}$ or 1.5

Chapter 7, Form A

1. D **2.** A **3.** D **4.** D **5.** D
6. C **7.** A **8.** A **9.** E

Chapter 7, Form B

1. A **2.** D **3.** B **4.** C **5.** A
6. C **7.** C **8.** A **9.** B **10.** B

Chapter 8, Form A

1. E **2.** B **3.** E **4.** D

Chapter 8, Form B

1. 52.5 **2.** 144 **3.** 30 **4.** 12 **5.** 25
6. 7 **7.** 60 **8.** 1 **9.** 240 **10.** 10
11. 4 **12.** 22

Chapter 9, Form A

1. C **2.** D **3.** D **4.** C **5.** C
6. C **7.** C **8.** A **9.** E

Chapter 9, Form B

1. A **2.** D **3.** D **4.** A **5.** C
6. B **7.** B **8.** D **9.** B **10.** D

Chapter 10, Form A

1. A **2.** E **3.** C **4.** A **5.** B
6. D **7.** B **8.** D

Chapter 10, Form B

1. 1 **2.** 6 **3.** 20 **4.** 2 **5.** 144
6. $\frac{2}{9}$ **7.** 17.1 **8.** 1 or 2
9. $\frac{1}{7}$ **10.** 8 **11.** 9

Chapter 11, Form A

1. E **2.** C **3.** C **4.** E **5.** E
6. B

Chapter 11, Form B

1. D **2.** B **3.** C **4.** A **5.** A
6. B **7.** C **8.** B **9.** A

Chapter 12, Form A

1. D **2.** E **3.** E **4.** B **5.** D
6. B **7.** C

Chapter 12, Form B

1. 12 or 30 **2.** $\frac{3}{2}$ **3.** 8
4. 3 **5.** 1.98 **6.** 32
7. 7 or 13 **8.** $\frac{5}{3}$

Chapter 13, Form A

1. A **2.** A **3.** C **4.** B **5.** E
6. E **7.** A

Chapter 13, Form B

1. D **2.** B **3.** B **4.** A **5.** A
6. A

Chapter 14, Form A

1. D **2.** B **3.** E **4.** E **5.** E

Chapter 14, Form B

1. 1630 **2.** 45 **3.** 72.5 **4.** 7.5
5. 300 **6.** 12 **7.** 24 **8.** 8
9. 3 **10.** 2 **11.** 1.25

Chapter 15, Form A

1. C **2.** C **3.** D **4.** A **5.** B
6. C

Chapter 15, Form B

1. B **2.** C **3.** B **4.** C **5.** B
6. D